T0271148

# Design of High-Performance Pre-Engineered Steel Concrete Composite Beams for Sustainable Construction

This accessible and practical shortform book details the properties and advantages of high-performance pre-engineered steel-concrete composite beams (HPCBs) for improving the sustainability of construction techniques. It also explains the analysis methods for testing HPCB systems.

The authors describe a new HPCB system that has been developed to reduce the input of raw materials and embodied $CO_2$ commonly associated with heavily loaded and long-spanned industrial buildings (which predominately comprise reinforced concrete) and improve the sustainability of the construction process. They provide several resources throughout to facilitate adoption by professionals. Design equations derived from Eurocode 4 approach for ultimate limit state and serviceability limit state and worked examples are included throughout. The authors discuss the feasibility for both materials and the full-scale beams and $CO_2$ reduction methods, including use of recycled concrete aggregate, ground granulated blast-furnace and silica fume to replace natural coarse aggregates and Ordinary Portland Cement. Guidance for testing HPCBs—including setup, test procedure and data collection and interpretation—is also given. The authors also elaborate on recommendations for finite element analysis for HPCBs. Design examples are appended to illustrate typical current practice using a 12 × 12 m grid floor with live load of 15 kPa. Various considerations for different parameters such as fire resistance are discussed. Finally, the authors present a case study of a recently completed industrial building in Singapore to quantify the benefits of using HPCBs over reinforced concrete and conventional composite construction.

Structural engineering professionals, whose work relates to long-span and heavy-loading industrial or commercial buildings, will benefit from the detailed guidance and focus on practical applications provided throughout this book. Postgraduate students of advanced steel and composite structures will also benefit from these descriptions.

**Ming-Shan Zhao** is Assistant Professor of Civil Engineering at the Singapore Institute of Technology.

**Sing-Ping Chiew** is Professor and Programme Leader of Civil Engineering at the Singapore Institute of Technology.

**Guan-Feng Chua** is a Doctorate of Engineering candidate at the Singapore Institute of Technology.

**Miao Ding** is a research engineer at the Singapore Institute of Technology.

**Yi Yang** is a senior engineer at JTC Corporation, Singapore.

**Zhengxia Cong** is Senior Director of Engineering (Building) at Woh Hup (Private) Limited, Professional Engineer in Civil Engineering and Special Professional Engineer in Geotechnical Engineering (Singapore), and Fellow of Charted Structural Engineer (UK).

# Design of High-Performance Pre-Engineered Steel Concrete Composite Beams for Sustainable Construction

Ming-Shan Zhao, Sing-Ping Chiew,
Guan-Feng Chua, Miao Ding,
Yi Yang, and Zhengxia Cong

**CRC Press**
Taylor & Francis Group
Boca Raton London New York

CRC Press is an imprint of the
Taylor & Francis Group, an **informa** business

First edition published 2025
by CRC Press
2385 NW Executive Center Drive, Suite 320, Boca Raton FL 33431

and by CRC Press
4 Park Square, Milton Park, Abingdon, Oxon, OX14 4RN

*CRC Press is an imprint of Taylor & Francis Group, LLC*

© 2025 Ming-Shan Zhao, Sing-Ping Chiew, Guan-Feng Chua, Miao Ding, Yi Yang and Zhengxia Cong

ISBN: 978-1-032-62691-8 (hbk)
ISBN: 978-1-032-62692-5 (pbk)
ISBN: 978-1-032-62693-2 (ebk)

DOI: 10.1201/9781032626932

Typeset in Minion
by SPi Technologies India Pvt Ltd (Straive)

# Contents

# Foreword

WITH AN EVER-INCREASING POPULATION, URBANIZATION AND economic development around the world, the built environment has emerged as a major contributor to climate change. Already now, construction materials such as steel, sand, gravel and limestone are now responsible for more than 50% of global resource consumption and their uses are projected to be more than double by 2060. To help mitigate climate change, sustainability in the built environment is gaining importance. Other than the use of alternative/recycled materials to replace construction materials with high carbon footprints and adopt low embodied carbon products, another promising approach is to optimize the design to reduce the materials used.

In Singapore, multi-storey industrial buildings are predominantly reinforced concrete (RC) structures designed for heavy loadings and long spans. While RC structural systems have served the industry well as a technically viable and economical solution, the deployment of large amount of low strength-to-weight ratio materials has resulted in structural members of huge sizes as well as high embodied carbon per gross floor area. From the sustainability point of view, resource and waste optimization by design are becoming increasingly important to meet the rising bar of sustainable construction. We must avoid over-design of structural elements by rationalizing the design live loads but without affecting building resilience and future adaptability.

To optimize material utilization and structural efficiency, the concept of high performance pre-engineered steel concrete composite beam (HPCB) is developed. The HPCB combines a high-performance green concrete slab with an asymmetrical high-performance steel section, which is designed to match the bending moments and shear forces for each beam of the actual building. The design is essentially based on EC4 with special consideration given to the resistance calculations under the ultimate limit state and serviceability requirements for industrial buildings.

This book is written primarily for structural engineers and designers who are familiar with basic EC4 design and wish to take advantage of high-performance materials to design more cost-competitive, efficient and sustainable long-span and heavy-loading buildings. This book includes feasibility studies for materials and connections and introduces guidance on the analysis methods and design equations that are validated with testing and finite element simulation. A case study of a recently completed industrial building in Singapore was also presented. We hope to ease the adoption of HPCB systems for professionals and proliferate their adoption for future buildings.

**Calvin Chung**
Assistant CEO,
Engineering & Operations Group
JTC Corporation

# Acknowledgements

THE AUTHORS APPRECIATE THE FINANCIAL AND TECHNICAL support from JTC Corporation throughout the course of the research project. The authors would also like to thank Mr Chee Kiat Tan (former Group Director of Engineering & Chief Sustainability Officer), Mr John Kiong, Er Kok Mun Lum and Mr Kian Wee Ng of JTC Corporation and Dr Yiaw Heong Ng of TTJ Design & Engineering Pte Ltd for their technical support and advice provided during the development of this research project and case studies.

# CHAPTER 1

# Introduction

## 1.1 SUSTAINABLE CONSTRUCTION OF NEXT-GENERATION INDUSTRIAL BUILDINGS

On the 2020 baseline, building and construction contribute to approximately 39% of the global carbon emissions, and 11% of this comes from the embodied carbon ($CO_2e$) associated with the materials and construction processes [1]. Already now, construction materials such as steel, sand, gravel and limestone are now responsible for more than 50% of global resource consumption, and their uses are projected to be more than doubled by 2060 [2]. To help mitigate climate change risks, sustainability in the built environment is gaining importance. It has been said that the world should target 40% reduction of embodied carbon by 2030 and net zero by 2050 [1]. However, currently there is no single material that can meet this target in view of performance, availability and productivity [3].

Industrial buildings in Singapore, which are predominately reinforced concrete (RC) structures, are heavily loaded (15–20 kPa) and long spanned ($\geq$12 m). While the RC solution is proven

DOI: 10.1201/9781032626932-1

to be still technically viable and economical, the low strength-to-weight ratio requires huge sizes for structural members, such as flooring systems, primary beams, columns and foundation elements. For example, a recently completed 9-storey RC warehouse (Figure 1.1) adopted heavily reinforced precast inverted tee beam with closely spaced shear links as primary beams (approximately 5,500 kg/m run), and doubly reinforced precast "single tee beam-slab" elements as secondary beams (approximately 2650 kg/m run), as shown in Figure 1.2. Besides, due to the high self-weight, large columns (e.g., 1,200 mm by 1,800 mm) have to be in place. This inevitably resulted in an enormous consumption of construction materials and eventually high $CO_2e$. It is estimated that this structural system is producing approximately 967.1 kg $CO_2e$ per gross floor area (GFA) for structural components alone. Note that the current maximum for $CO_2e$, as established by the regulatory body in Singapore, is 1,000 kg per GFA for buildings, encompassing both architectural and structural components. It is therefore urgently needed to develop an alternative high-performance flooring system that requires less input of raw materials and eventually reduces $CO_2e$.

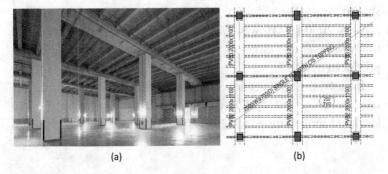

(a)                                        (b)

FIGURE 1.1 Next-generation RC industrial buildings in Singapore featuring high rise, long span and heavy loading. (a) Recently completed 9-storey RC warehouse; and (b) Typical floor plan.

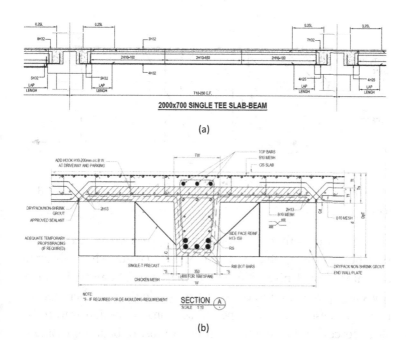

**FIGURE 1.2** Doubly reinforced T beams as the secondary beams. (a) Elevation view; and (b) Cross-section view.

## 1.2 DESIGN TO THE EUROCODES: MATERIAL UTILIZATION AND STRUCTURAL EFFICIENCY

Concrete is known for its high compressive strength and weak tensile resistance. In RC beam design, steel reinforcement bars (rebars) are used to take tension assuming concrete has negligible tensile strength. According to EN 1992, the neutral axis is limited to 0.45d, where "d" is the depth from the extreme concrete fibre in compression to the centroid of the reinforcement bars [4]. This ensures the strain compatibility of concrete and reinforcement, and a "balanced section" to achieve a ductile failure. However, it also inevitably increases the size of beams and ultimately leads to an increase in $CO_2e$, as not less than 55% of the concrete materials are not contributing to the design moment resistance, as shown in Figure 1.3.

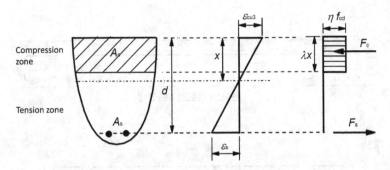

FIGURE 1.3   Rectangular stress distribution for RC beams subjected to bending.

Steel, on the other hand, has excellent resistance against tension, but it is susceptible to various buckling issues when subjected to compression. Universal beams (UB) that are of standard dimensions are widely used in construction worldwide for their convenience in design. Webs of the UB sections are designed to resist shear forces while flanges are to resist most of bending moment experienced by the beams. UBs are proven to be reliable and efficient as a general beam solution to common steel structures. To mitigate the local buckling issues, all 64 numbers of UBs in grades S275 & S355 and ranging from 30 kg/m to 273 kg/m subjected to bending are of either Class 1 or Class 2, as shown in Table 1.1.

TABLE 1.1   Classification of universal steel sections for standalone steel Structures

| Grade | Section | Web Subject to Bending | Web Subject to Compression | Flange Subject to Compression |
|---|---|---|---|---|
| S275 | UB | All Class 1 or 2 | 24 Class 1 or 2 43 Class 3 | All Class 1 or 2 |
|  | UC | All Class 1 or 2 | All Class 1 or 2 | All Class 1 or 2 |
| S355 | UB | All Class 1 or 2 | 14 Class 1 or 2 53 Class 3 | All Class 1 or 2 |
|  | UC | All Class 1 or 2 | All Class 1 or 2 | 2 Class 3 |

*Note:*
- This summary includes 67 UB sections from 30 kg/m to 273 kg/m.
- This summary includes 36 UC sections from 23 kg/m to 634 kg/m.

When higher loading capacity is required, structural steel sections and reinforced concrete slabs are commonly combined to form composite beams, as shown in Figure 1.4. Such steel concrete composite beams are more efficient and productive compared to either material acting alone. In the ideal situation where composite action is fully achieved by a sufficient number of shear connectors, the steel beam should be taking tensile stresses and the concrete slab should primarily take compressive stresses, thus utilizing the favourable attributes of each material in an efficient manner (Figure 1.4). Furthermore, the concrete slab also prevents lateral torsional buckling of the steel beam, which is a common phenomenon in an unrestrained steel beam.

However, as shown in Figure 1.5, it is clear in the design stress block that the bottom portion of concrete slab and the upper

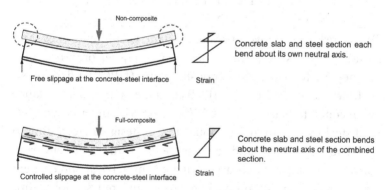

FIGURE 1.4    From EC2 and EC3 to EC4: potential to enhance structural efficiency.

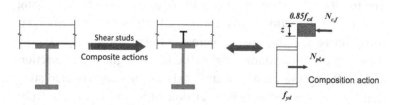

FIGURE 1.5    Stress block of steel-concrete composite beam subjected to bending moment.

flange of steel are too close to the plastic neutral axis to contribute significantly to the bending moment resistance, but result in a substantial amount of redundant weight. For example, in the case of a 762 × 267 $UB$173, the top flange alone can contribute to redundant weight of 45.3 kg/m (26.2% the total weight of the UB) if used in a steel concrete composite beam. On the other hand, conventional UB sections come with fixed dimensions. While standard dimensions provide convenience in design and estimation, the thickness, width and height are in English units and prohibit customization. Such inflexibility and limited sizes of UB available inevitably end up with structural inefficiency, which is expressed as redundant weight and $CO_2$e.

## 1.3 CONCEPT OF HIGH-PERFORMANCE PRE-ENGINEERED STEEL CONCRETE COMPOSITE BEAMS

To address the above-mentioned structural inefficiency and sustainability issues, the concept of high-performance pre-engineered steel concrete composite beams (HPCB) is developed, as shown in Figure 1.6. The HPCB combines a high-performance green concrete slab in grade C55/67 or C60/75 and an asymmetrical steel section in grade S460M that is designed according to the bending moment and shear force diagrams. Different from conventional universal beams that are symmetrical and efficient for standalone applications, the pre-engineered section minimizes the size of the top flange and web, which contribute little to bending moment resistance in a composite beam (Figure 1.5). The top flange is reduced in size to support the profiled decking and shear connection studs, while the bottom flange is enlarged to optimize the composite beam's moment capacity. Not only reducing the redundant weight, the use of lighter steel section also shifts up the plastic neutral axis to the concrete slab such that the steel is subjected to tension only and thus free from buckling issues.

## 1.4 BENEFITS OF USING HPCB

The HPCB usage would not only benefit the environment in terms of sustainability but also provide economic benefit in terms of costing. The incorporation of high-performance and pre-engineering materials significantly reduced the embodied carbon ($CO_2e$). In a case study performed, the HPCB reduced the $CO_2e$ of the next-generation RC structure by approximately 30%, and in addition, the usage of total materials was also significantly reduced by approximately 60%. As the mass of each structure element is reduced, a lower capacity of lifting equipment can therefore be deployed, further reducing carbon emission associated. Owing to the significantly reduced material input, the overall cost to construct the next-generation industrial building using HPCB solution is expected to be 5–10% cheaper than the conventional RC solution.

Another benefit of using HPCB would be the reduction in construction duration. It is estimated that 28 man-days would be required for the construction of a single 12 by 12 m grid for RC structure, while 24 man-days would be sufficient for the same 12 by 12 m layout adopting HPCB solution (More details can be found in the Appendix B: Case Study). This not only increases the productivity and efficiency in construction stage but also brings in economic benefits by reducing the operation cost related.

## 1.5 ABOUT THIS BOOK

A series of design guides on high-strength steel concrete composite beams are readily available in literature with a focus on high capacity and design with symmetrical universal sections. The niche of this book is the design and validation of HPCBs made with high-performance green concrete and pre-engineered steel sections. HPCB is a novel solution especially suitable for heavily loaded and long-spanned industrial buildings. While the concept is still within Eurocode 4, the industry always faces critical questions when it comes to application such as how to design for

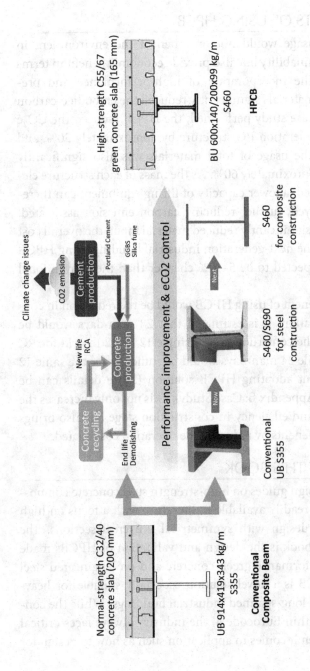

FIGURE 1.6  Concept of HPCB to achieve greater sustainability.

ultimate limit state and serviceability limit state, how to ensure the performance of materials when multiple "green" ingredients are mixed, and where is the evidence or track record for validating the overall beam behaviour?

This book contains comprehensive knowledge about the HPCB, including design, feasibility for materials, testing and validation of the full-scale beams, worked examples and a case study for productivity, cost and carbon analysis. The book is expected to endow structural engineers with the confidence to adopt HPCB in a safe and efficient manner in design and construction.

## REFERENCES

1. WorldGBC, Advancing Net Zero Status Report 2022 World Green Building Council (2022).
2. OECD, *Global Material Resources Outlook to 2060: Economic drivers and environmental consequences* OECD Publishing, Paris (2019).
3. Blanco, J.L., Engel, H., Imhorst, F., Ribeirinho, M.J., Sjodin, E., *Seizing the decarbonization opportunity in construction*, Engineering, Construction & Building Materials, McKinsey & Company (2021).
4. EN 1992-1-1 Eurocode 2: *Design of concrete structures – Part 1-1: General rules and rules for buildings* (2005).

# CHAPTER 2

# Materials and Buildability

## 2.1 HIGH-PERFORMANCE GREEN CONCRETE

Urbanization and economic development increase the demand for new buildings and infrastructure and hence for concrete. Already now, concrete is the second most used material by mass after water [1]. Even though concrete has a relatively low specific $CO_2$ emission of below 200 kg/ton, its abundance makes it responsible for about 11% man-made $CO_2$ emissions [2], which has become a major contributor to climate change. Norway's Center for International Climate Research reported that in 2021, nearly 2.9 billion tonnes of $CO_2$ were emitted from the worldwide production of cement for buildings, roads and other infrastructure. This $CO_2$ emission amount was doubled as compared to cement emissions in 2002 [3].

While the design of HPCB focuses on the optimization of geometry through pre-engineering efforts, the adoption of higher-performance materials could apparently synergize with this approach and result in greater benefits in sustainability. EN 206-1 defines high-strength concrete as concrete with a compressive

DOI: 10.1201/9781032626932-2

strength class higher than C50/60 in the case of normal-weight concrete. Although EC4 limits the use of high-strength concrete to C60/75 and below, significant weight saving is still achievable if the strength of concrete can be increased, for example, from the most commonly used C30/37 to C60/75, as shown in Table 2.1.

Conventional high-performance concrete targeting superior mechanical properties is typically produced using high-quality aggregates, a low water-to-cement ratio or water-to-binder ratio, and a higher content of Portland cement [4]. Figure 2.1 shows the composition of $CO_2$e of a typical ready-mixed concrete in grade C32/40 reinforced with 2% steel reinforcement (rebar). The $CO_2$e of each gradient is taken from the ICE embodied carbon database [5]. It can be seen from Figure 2.1 that rebar and cement contribute to 95% of the total $CO_2$e in this case. Therefore, it is not surprising that high-performance concrete emits a relatively higher carbon emission than normal-strength concrete as more cement is required for its production [6]. However, the overall carbon emission may not be necessarily higher as the volume of high-performance concrete required is significantly lower than normal-strength concrete for construction of the same building. This result is consistent with the research conducted by Habert et al., that the environmental impact is significantly reduced, by an average of 20% when conventional concrete is replaced with high-performance concrete for bridge construction [7].

TABLE 2.1    Strength classes for normal-weight concrete in EC2

|  | Normal-Strength Concrete | | | | High-Strength Concrete | | |
|---|---|---|---|---|---|---|---|
|  | C30/37 | C35/45 | C40/50 | C45/55 | C50/60 | C55/67 | C60/75 |
| $f_{ck}$ (MPa) | 30 | 35 | 40 | 45 | 56 | 55 | 60 |
| $f_{ck, cube}$ (Mpa) | 37 | 45 | 50 | 55 | 60 | 67 | 75 |
| $f_{ctm}$ (Mpa) | 2.9 | 3.2 | 3.5 | 3.8 | 4.1 | 4.2 | 4.4 |
| $E_{cm}$ (Mpa) | 33 | 34 | 35 | 36 | 37 | 38 | 39 |

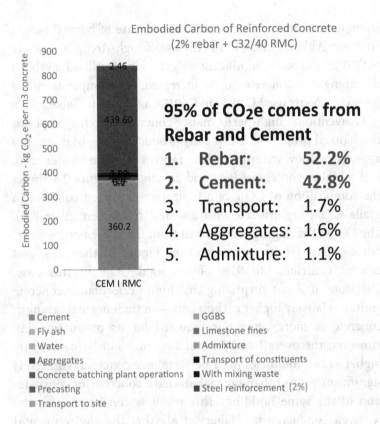

FIGURE 2.1   Typical $CO_2e$ of C32/40 precast reinforced concrete.

The interest in developing high-performance green concrete (HPGC) that partially replaces the virgin materials in production of conventional high-performance concrete is growing rapidly [8], particularly in the use of various waste/by-product materials as either binder or/and aggregates [9].

## 2.1.1  Use of Alternative Materials (Binders and Aggregates) for Concrete Production

### 2.1.1.1  Reducing $CO_2e$ with the Use of SCMs to Replace OPC

As most of concrete's emissions result from the production of Ordinary Portland cement (OPC) clinker, researchers and cement producers have put lots of effort into lowering the clinker content in cement and improving the efficiency of clinker production. Supplementary cementitious materials (SCMs) which possess pozzolanic and/or cementitious properties are often used in conjunction with OPC for concrete production to reduce carbon emissions [6, 10]. Extensive works have been conducted on the hydration and reaction mechanisms of common SCMs such as fly ash, ground granulated blast-furnace slag (GGBS), silica fume and fine limestone in cementitious materials and their applications in geotechnical and building infrastructure. Considering the vast availability and high requirement for workability and strength in high-performance concrete, attention is paid to GGBS and silica fume in this book.

- GGBS is a by-product of iron- and steel-making process using blast-furnace, obtained by quenching molten iron slag in water or steam, to produce a glassy, granular product that is then dried and ground into a fine powder. GGBS is known to possess both cementitious and pozzolanic properties and is allowed to replace up to 85% OPC in CEM III cement [6]. Its main advantage is its slow release of hydration heat, allowing limitation of the temperature increase in massive concrete components and structures during cement setting and concrete curing. However, a high percentage of GGBS replacement is likely to adversely affect the early compressive strength of concrete, which is often of concern for structural usage.

- Silica fume is a by-product of silicon and ferrosilicon alloy production. It is an extremely fine spherical particle of amorphous $SiO_2$ with an average particle diameter of approximately 150 nm that formed from escaping gaseous through the oxidization and condensation of $SiO_2$ [7]. The main field of application is as pozzolanic material for high-performance concrete at replacement levels between 3% and 10% for OPC, for its benefits in improving mechanical properties, particularly compressive strength, bond strength and abrasion resistance, due to the pozzolanic reaction between silica fume and free calcium hydroxide [11]. Replacement levels above 10% can lead to further durability improvements, but the workability of the concrete can be problematic.

Typical chemical composition of the SCMs, in comparison to OPC, is shown in Table 2.2.

As shown in Figure 2.2, the Building Embodied Carbon Calculator (BECC) developed by the Singapore Green Building Council [12] suggests emission factors of 0.1 kg-$CO_2$e/kg for GGBS and 0.08 kg-$CO_2$e/kg for fly ash, which are significantly lower than 0.95 kg-$CO_2$e/kg for OPC. Although the $CO_2$e values of SCMs vary according to their source of manufacturing, it is widely acknowledged that using SCMs to replace OPC is a highly effective approach to reduce global carbon emissions [13]. However, it shall be noted that the performance of different cement mixtures, especially the early-age mechanical properties, is highly dependent on the characteristics of the SCMs, cement composition and the targeted application [14]. A thorough understanding of the properties of SCMs and their reaction mechanisms within the

TABLE 2.2  Typical chemical composition of common SCMs

| (%) | OPC | GGBS | Coal FA(low Ca) | Silica Fume | Fine Limestone |
|---|---|---|---|---|---|
| $SiO_2$ | 19.8 | 35.4 | 65.8 | 94.0 | — |
| CaO | 63.2 | 40.5 | 1.8 | 0.4 | — |
| $Al_2O_3$ | 5.0 | 13.0 | 21.5 | 0.1 | — |
| $Fe_2O_3$ | 2.4 | 0.37 | 4.6 | 0.1 | — |
| MgO | 3.3 | 8.0 | 0.8 | 0.4 | <0.45 |
| $K_2O$ | 1.2 | — | 1.9 | 0.9 | — |
| $SO_3$ | 3.0 | 2.4 | 1.4 | 1.3 | — |
| $Na_2O$ | 0.1 | — | 0.1 | 0.1 | — |
| $TiO_2$ | 0.3 | — | 0.7 | 0.3 | — |
| $CaCO_3$ | — | — | — | — | 99.0 |
| LoI | 2.5 | 0.88 | 1.4 | 4.7 | — |

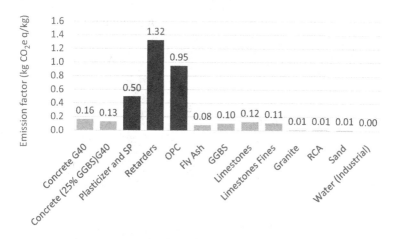

FIGURE 2.2   $CO_2e$ of common constituents of concrete.

specified HPGC mixture is crucial prior to their widespread adoption in real-world applications.

### 2.1.1.2 Recycling and Upcycling Demolition Waste for New Builds

It is widely acknowledged that improving circularity in construction or recycling demolition waste for new builds are effective approaches to improve sustainability. One of the most remarkable success stories would be the production of recycled rebars using 100% scrap metal through an electric arc furnace. Compared to rebars produced from natural resources (approximately 2.8 kg-$CO_2e$/kg), the recycled rebar could lower the $CO_2e$ to approximately 0.5 kg-$CO_2e$/kg, which could then significantly reduce the $CO_2e$ in construction of reinforced concrete structures (see Figure 2.1). Similarly, the use of secondary aggregates, such as recycled concrete aggregate (RCA), helps conserve natural resources [15], although the $CO_2e$ benefit through partial replacement of natural coarse aggregate with RCA seems insignificant, as shown in Figure 2.3. As stated in EN 206:2014 (Table 2.3), the replacement of coarse aggregate is limited by the different aggregate types and concrete exposure classes. In Singapore practice, the proportion

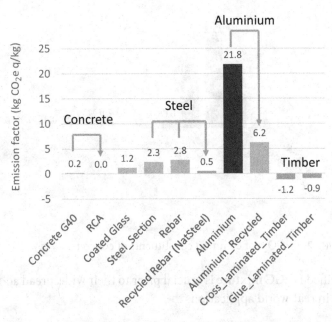

FIGURE 2.3  Recycling materials at their end-life for lower $CO_2e$ factors.

TABLE 2.3  Permitted RCA replacement (EN 206:2014)

| | Exposure Class | | | |
|---|---|---|---|---|
| RCA type | X0 (%) | XC1 & XC2 (%) | XC3, XC4, XF1, XA1 & XD1 (%) | All other (%) |
| Type A | 50 | 30 | 30 | 0 |
| Type B | 50 | 20 | 0 | 0 |

of replacement is limited to not more than a mass fraction of 20% of the coarse aggregate (SS 544: Part 1: 2019).

Despite the tempting benefits of sustainability, the use of RCA in structural concrete, especially in high-performance concrete, is often of concern because of its inferior properties compared to natural aggregates [17–19]. In an experimental investigation conducted by S.B Daneti, the effect of compressive strength of concrete is examined for three different ready-mixed concrete classes

(C32/40, C50/60 and C65/80) with each strength class having RCA replacement of 20, 50 and 100% [16]. It was reported that the influence of RCA on the compressive strength was no longer negligible if the grade was higher than C50/60.

The sequestration of $CO_2$ into RCA has demonstrated great potential in lowering the carbon footprint and enhancing the mechanical properties of overall concrete. The mechanism by which $CO_2$ transforms the calcium hydroxide $Ca(OH)_2$ into calcium carbonate $(CaCO_3)$ permits the filling of air voids providing the densification of a cement paste, which mitigates negative properties. Carbonation is a natural mechanism that can take years or even decades to achieve the required depth of penetration. The injection of a high concentration of $CO_2$ at pressure under suitable humidity and environment accelerates the chemical reaction and realizes an enhanced cement paste. The carbonation process is a complex process of chemical reactions and mass transfer. The Commonwealth Science and Industrial Research Organisation (CSIRO) suggests that $CO_2$ mineral carbonation consists of three distinctive steps:

a. Leaching of solid matrix

b. $CO_2$ dissolution into solution

c. Carbonate precipitation

According to Liang et al., the chemical reaction between $CO_2$ and RCA can be described as follows: $CO_2$ reacts with the hydration production in the adhered old mortar, Calcium Silicate Hydrate and hydrated lime, to produce Calcium Carbonate and silica gel, and likewise reacts with the unhydrated clinker to produce more Calcium Carbonate [17]. The chemical equations are as follows:

Reaction between $CO_2$ and $Ca(OH)_2$/C-S-H

$$Ca(OH)_2 + CO_2 \rightarrow CaCO_3 + H_2O \qquad (2.1)$$

$$C - S - H + CO_2 \rightarrow CaCO_3 + SiO_2 \bullet \mu H_2O \qquad (2.2)$$

Reaction between $CO_2$ and $C_3AH6$

$$4CaO \bullet Al_2O_3 \bullet 13H_2O + 14CO_2 \rightarrow \\ 4CaCO_3 + 2Al(OH)_3 + 10H_2O \qquad (2.3)$$

Reaction between $CO_2$ and unhydrated cement clinker

$$3CaO \bullet SiO_2 + 3CO_2 + \mu H_2O \rightarrow SiO_2 \bullet \mu H_2O + 3CaCO_3 \quad (2.4)$$

$$2CaO \bullet SiO_2 + 2CO_2 + \mu H_2O \rightarrow 2SiO_2 \bullet \mu H_2O + 2CaCO_3 \quad (2.5)$$

## 2.1.2 Feasibility Study of Economic HPGC

Experimental investigation was carried out to validate and codify the fresh and hardened mechanical properties of HPGC in grade C55/67 incorporated with multiple "green" gradients used to reduce the input of virgin materials. The synergy among SCMs including GGBS and silica fume and RCA/carbonated RCAs is also evaluated.

### 2.1.2.1 Materials and Design Mix

The cement used in the HPPGC was OPC CEM I with a specific gravity of 3.15. SCMs include blast-furnace cement CEM III/B with a specific gravity of 2.8–2.95 and Elkem Mircosilica® undensified 940U silica fume with a specific gravity of 2.2–2.3. These cementitious materials meet the specifications stated in EN197-1:2014, EN15167-1:2008 and EN 13263-1:2005+A1:2009, respectively. A batch of crushed granite with a maximum size of 20 mm and specific gravity of 2.65, was used as coarse aggregate, while concrete sand with a specific gravity of 2.55 was used as fine aggregate. Both coarse and fine aggregate complied with EN 12620, and the relevant standards listed in EN 12620. A batch of recycled concrete aggregate with maximum size of 20 mm was obtained from a recently demolished building in Singapore. The recycled concrete aggregate, $R_c90$, complied with the specification of EN 12620. To achieve adequate workability for this HPGC incorporated with various waste/by-product materials, Mapei® Dynamon NRG1030, was used as a high-range super-plasticiser. A summary of the materials used is shown in Table 2.4.

The BRE concrete mix design method is adopted to determine the water-to-binder ratio, aggregate-to-binder ratio, workability and the ratio of fine to coarse aggregate to produce the targeted HPGC in grade C55/67. The basic design mix is shown in Table 2.5. Note that besides the requirement for compressive strength, it is recommended to pay attention to the aggregate-to-cement ratio and the ratio of fine to coarse aggregates so as to achieve high workability for ready-mixed concrete, especially when multiple "green" gradients are used.

TABLE 2.4    Materials used for producing HPGC in grade C55/67

| Material | Code/Standard | Specific Gravity |
| --- | --- | --- |
| CEM I | EN 197-1:2014 | 3.15 |
| CEM III/B | EN 15167-1:2008 | 2.8–2.95 |
| Silica fume | EN 13263-1:2015+A1:2019 | 2.2–2.3 |
| Recycled concrete aggregate | EN 12620 | 2.35–2.4 |
| Coarse aggregate | EN 12620 | 2.65 |
| Fine aggregate | EN 12620 | 2.55 |
| Super-plasticiser | EN 934 | — |

TABLE 2.5    Basic design mix for HPGC C55/67

| Material | Quantity (kg/m$^3$) |
| --- | --- |
| Binder | 567 |
| Water-to-binder ratio | 0.256 |
| Fine aggregate | 564 |
| Coarse aggregate | 1,116 |
| Superplasticiser | 8.51 |

The adjustment of the design mix for parametric study includes using up to approximately 20% GGBS and 3.5% silica fume as SCMs to replace OPC and using up to 20% RCA and carbonated RCA to replace natural coarse aggregates. A comprehensive experimental matrix including single, double and tribble variables can be found in Table 2.6.

### 2.1.2.2 Carbonation of RCAs

To determine the effect of relative humidity on the $CO_2$ sequestration efficiency in carbonation of RCA, a pressure chamber setup shown in Figure 2.4 is used. Anhydrous silica gel with different moisture contents was placed in the bottom of the chamber to maintain the relative humidity at target levels. This carbonation technique adopted for the carbonation of RCA is similar to Zhan et al. [18].

- Firstly, oven-dried RCA is weighed before placing into the chamber, and a cylinder of pure $CO_2$ is used to inject $CO_2$ at 1 bar of pressure into the chamber to carbonate the RCA [18].

- Secondly, the RCA is collected at a pre-determined duration of carbonation and oven-dried at approximately 105°C for 24 hours to determine the mass gained from carbonation.

- Finally, after measuring of mass gained, the RCA is placed back into the chamber to continue the carbonation process. The process is repeated until a steady state of carbonation is observed.

Table 2.7 presents the different relative humidity experimented and Figure 2.5 represents the $CO_2$ uptake rate for S1 to S4, respectively. S1 to S3 have an aggregate size ranging from 10 mm to 20 mm, while S4 has an aggregate size ranging from 4 mm to 10 mm. Sample S3 has a total carbonation of 1.28% gained in mass under a relative humidity of 80.9, followed by S2 of close to 1% under a relative humidity of 46.9. As expected, the amount of $CO_2$

TABLE 2.6 Details of experimental matrix

| Mix | CEM I | CEM III/B | Silica Fume | kg/m³[%] CA (NA) | CA (RCA) | CA (CRCA) | FA | Water |
|---|---|---|---|---|---|---|---|---|
| Control | 567 | 0 | 0 | 1116 | 0 | 0 | 564 | 144.5 |
| R1 | 567 | 0 | 0 | 1004.4 | 111.6 [10] | 0 | 564 | 144.5 |
| R2 | 567 | 0 | 0 | 892.8 | 223.2 [20] | 0 | 564 | 144.5 |
| CR1 | 567 | 0 | 0 | 1004.4 | 0 | 111.6 [10][a] | 564 | 144.5 |
| CR2A | 567 | 0 | 0 | 892.8 | 0 | 223.2 [20][a] | 564 | 144.5 |
| CR2B | 567 | 0 | 0 | 892.8 | 0 | 223.2 [20][b] | 564 | 144.5 |
| G1 | 410.6 | 156.4 | 0 | 1116 | 0 | 0 | 564 | 144.5 |
| G2 | 410.6 | 156.4 | 0 | 1116 | 223.2 [20] | 0 | 564 | 144.5 |
| G3 | 410.6 | 156.4 | 0 | 1116 | 0 | 223.2 [20][b] | 564 | 144.5 |
| S1 | 390.1 | 156.4 | 19.5 | 892.8 | 223.2 [20] | 0 | 564 | 144.5 |
| S2 | 390.1 | 156.4 | 19.5 | 892.8 | 0 | 223.2 [20][b] | 564 | 144.5 |

*Notes:*
[a] 0.13% carbonation intake used, based on mass change
[b] 1.3% carbonation intake used, based on mass change

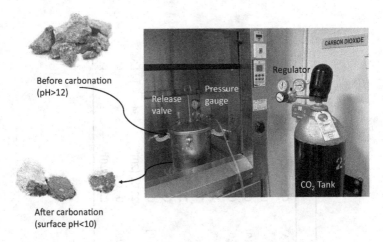

FIGURE 2.4    Lab setup for accelerated $CO_2$ sequestration in RCA.

TABLE 2.7    Humidity and temperature of carbonation process

| Sample | Relative Humidity (RH) | Temperature (°C) |
| --- | --- | --- |
| S1 | 19.5 | 23.4 |
| S2 | 46.9 | 33.2 |
| S3 | 80.9 | 36.8 |
| S4 | 17.0 | 25.5 |

adsorbed by the RCA was higher in the high-humidity environment as compared to the RCA in the drier environments. This is due to the chemical reaction (mentioned in Section 2.1.1.2) requiring $CO_2$ to be dissolved into aqueous form before carbonate precipitation can take place. It is observed that despite having a slightly lower relative humidity of 17.0 for S4 and 19.5 for S1, S4 has a higher mass change, 0.58%, as compared to S1 of 0.16%. This higher absorption of $CO_2$ can be attributed to the increase of surface area when the size of aggregate decreases, which allows for more "contact" surface for carbonate precipitation. The relative humidity plays a significant role in the amount of carbonation. However, by contrast, it seems to have negligible influence on the

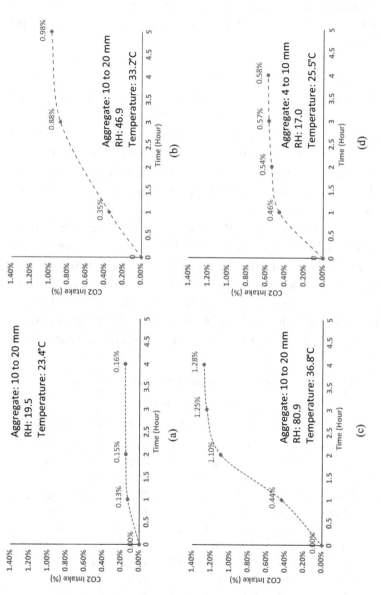

FIGURE 2.5  CO$_2$ uptake rate. (a) Sample S1; (b) Sample S2; (c) Sample S3; and (d) Sample S4.

total duration of carbonation. All samples exhibited a similar absorption trend, reaching a steady state at the fourth hour, despite an exponential climb during the first or even the second hour.

### 2.1.2.3 Compressive Strength

Table 2.8 presents the compressive strengths of all the specimens tested (listed in the experimental matrix in Table 2.5). As can be seen from Table 2.8, all the design mixes except for R2 (20% RCA) achieved compressive strength of 67 Mpa with a small standard deviation. The design mix S1 was selected as the concrete materials for the subsequent HPCB experimental studies, for its optimal strength, workability and low estimated $CO_2$e.

The compressive strengths of the specimens were within expectation, aside from specimens G1 and S2. The ultimate compressive strength decreased with the introduction of RCA, with compressive strength decreasing about 5% for 10% of RCA replacement and 7% for 20% of RCA replacement. In contrast, the addition of SCM increased the compressive strength despite the RCA replacement. It is observed that a 9% increase with the introduction of

TABLE 2.8 Compressive strengths of the tested design mixes

| | Compressive Strength (Mpa) | | | | | | |
|---|---|---|---|---|---|---|---|
| **Mix** | **Day 3** | **SD** | **Day 7** | **SD** | **Day 28** | **SD** | **$CO_2$e (kg)** |
| Control | 52.1 | 0.6 | 61.8 | 0.1 | 70.7 | 1.8 | 539 |
| R1 | 55.1 | 3.5 | 60.3 | 2.6 | 67.2 | 1.2 | 539 |
| R2 | 44.3 | 3.9 | 61.5 | 1 | 65.7 | 1.4 | 539 |
| CR1 | 62.8 | 3.7 | 74.4 | 4.5 | 77.4 | 2.0 | 539 |
| CR2A | 46.6 | 1.5 | 62.0 | 0.2 | 67.6 | 7.0 | 539 |
| CR2B | 53.0 | 4.3 | 65.1 | 4.6 | 74.8 | 1.9 | 539 |
| G1 | — | | 70.9 | 5.6 | 77.1 | 0.9 | 443 |
| G2 | 53.0 | 0.7 | 65.1 | 3.7 | 74.8 | 1.5 | 443 |
| G3 | 46.6 | 3 | 62.0 | 0.3 | 67.6 | 1.6 | 443 |
| S1 | 52.2 | 0.7 | 61.9 | 0.3 | 75.6 | 1.6 | 418 |
| S2 | 42.8 | 0.1 | 54.1 | 1.7 | 68.5 | 2.7 | 418 |

GGBS (G1) to control, and a 13.8% increase from R2 to G2. The decrease in compressive strength when RCA is introduced to the concrete can be attributed to the inherent properties that are inferior when compared to natural aggregate. According to Feng et al. [26], the old cement paste attached to the RCA has micro-cracks and voids, and this caused a loose and low micro-hardness of interfacial transition zone (ITZ), leading to a poorer performance of concrete produced. On the other hand, the addition of SMC increases compressive strength, which can be attributed to the increase in the hydration activities that lead to an increase of Calcium Silicate Hydrate and Calcium Hydroxide.

Similarly, when higher percentage of CRCA was introduced to replace the natural aggregate, ultimate compressive strength was observed to decrease as well. The decrease of compressive strength for CRCA is less significant as compared to non-CRCA, with a reduction from 77.4 MPa to 67.4 MPa. However, specimen CR1 has exhibited a higher early and ultimate compressive strength as compared to the control specimen, exhibiting an increase of 6.7 MPa which is equivalent to 10%. It is observed that compressive strength increases with the increase of carbonation concentration by 6.4 MPa, 3.1 MPa and 7.2 MPa for 3rd, 7th and 28th day respectively. Furthermore, the higher concentration of carbonation has also exhibited an improvement in compressive strength when compared to the control mix by 0.9 MPa, 3.3 MPa and 4.1 MPa for 3rd, 7th and 28th day, respectively. This improvement in compressive strength when CRCA is used for replacement, instead of non-CRCA, can be attributed to the formation of nano calcium carbonate. This formation of calcium carbonate improves the ITZ, creating a denser ITZ when compared to non-carbonate RCA [18]. As carbonation of RCA reduces the water absorption, the thickness of the water film reduces, therefore, improving the ITZ [20]. It is also reported by Luo et al. that carbonation reduces the ITZ width and improves the strength of the ITZ. Furthermore, Li et al. have reported that crack size was reduced.

It is observed that the RCA replacement reduces the compressive strength. When GGBS is incorporated into the mix with RCA, the compressive strength shows an estimated 3% increment. The inclusion of SF with GGBS and RCA shows the greatest compressive strength of 75.6 MPa, an approximate of 15% increment when compared to RCA mix and 12.7% improvement to GGBS and RCA mix. The significant increase in compressive strength with the inclusion of SF can be attributed to the pozzolanic activities that support the formation of additional calcium silicate hydrates in the RCA [21].

When CRCA and SCMs were used individually in the concrete, the compressive strength was observed to increase. An increase in compressive strength was expected. However, on the contrary, the compressive strength has not increased but reduced. It was observed that GGBS with CRCA (Specimen G3) registered a drop in compressive strength from 77.1 MPa to 62.3 MPa (19.2% drop), as compared to specimen G1. The introduction of SF did slightly increase the concrete strength to 68.5 MPa. However, this is a reduction of 7.1 MPa from specimen S1 and 1.6 MPa from control.

There is currently no strong evidence of this phenomenal occurrence. A possible hypothesis towards this phenomenal scenario could be that the pozzolanic materials replaced the cement and promoted secondary hydration reactions; its addition would replace the alkalinity of RCA and the content of calcium hydroxide that is the buffer for external erosion, therefore resulting in a larger carbonation depth.

## 2.2 FABRICATION OF HIGH-PERFORMANCE STRUCTURAL STEEL S460M SECTIONS

### 2.2.1 Steel Material—TMCP S460M

Structural steel is one of the most popular materials employed in civil engineering construction due to its high strength, stiffness, toughness and ductile properties. With the continuous development of new design ideas and manufacturing technologies, steel as a material for structural application has undergone significant changes. In the 1900s, most primary structural steel only had nominal yield strengths of about 220 MPa, which is equivalent to today's "mild steel". The once so-called "high-strength" steel S355 is now a widely used structural material. Owing to higher strength-to-weight ratio and potential to offer a lower total material cost and aesthetic advantages, the interest in using high-strength steel with a minimum yield strength of 460 MPa in steel-concrete composite structure applications has been increasing in the past decade.

Essentially, steels are just alloys of iron with carbon, which may contribute up to 2.1% of its weight. Although the properties of steel are greatly affected by its chemical composition, various treatments to which the steel may be subjected after leaving the manufacturing line can still remarkably affect the mechanical properties. The main reason that high-strength steel and normal-strength steel have similar chemical composition yet drastic different mechanical properties except for elastic modulus is that high-strength steel usually goes through heat treatment hardening processes, more specifically, quenching and tempering, and the thermo-mechanical controlled rolling process (TMCP). The former is more common for S690, while the latter has become popular in the grade of S460M for its cost-effectiveness.

The concept of TMCP combines controlled hot rolling with accelerated cooling to control the microstructure of the materials. The microstructure provides the "fingerprints" of a steel product that determine its properties. As a result, TMCP enables the production of as-rolled steels with final properties that are tailored to the requirements and specifications of a particular application. The goal of the TMCP is to produce cost-efficient steel strips and plates with properties required for a specific application. In addition to strength, the hardness and toughness, weldability and corrosion resistance of the steel are usually the targeted material features of the TMCP. The TMCP grade S460M steel has a minimum yield strength of 460 MPa, a tensile strength between 540 MPa and 720 MPa, and minimum elongation of 17%.

## 2.2.2 Welding Procedure Specification (WPS) for S460M

The HPCB utilizes structural steel grade of S460M for the pre-engineered section, where welding is a critical process for fabrication. A WPS for production welding, repairing welding and built-up welding must be qualified by welding procedure tests complying with EN ISP 15614-1. Furthermore, the welding procedure tests for full penetration butt weld shall be carried out in accordance with EN ISO 15614 and EN ISO 15613.

Standard butt joints for welding procedure test were fabricated by joining two pieces of 150 mm × 300 mm × 25 mm S460M plates by submerged arc welding (SAW) technique, as shown in Figure 2.6. Pre-heat and interpose temperatures were maintained at approximately 65 °C. Heat input per pass for SAW varied from 2.1 kJ/mm to 2.7 kJ/mm.

In Tables 2.9 and 2.10, the non-destructive and destructive testing respectively and the relevant standards and test results are listed. Specimens for the all-weld tensile test, transverse weld tensile test, bend and rebend test, and Charpy V-notch impact test (tested at −20°C) are shown in Figure 2.7. As specified in EN 1993-1-8, the resistance of a full penetration butt weld should be equal to or higher than the design resistance of the weaker of the

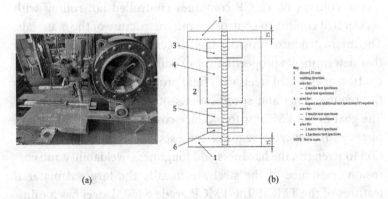

(a)                                        (b)

FIGURE 2.6  Welding procedure specification tests for full penetration butt weld. (a) Trial welding for S460M; and (b) Specimens to be machined for testing.

TABLE 2.9    Non-destructive testing

| Type of Test | Extend of Testing (%) | Test Standard |
|---|---|---|
| Visual Testing | 100 | ISO 17637 |
| Radiographic or Ultrasonic Test | 100 | ISO 17636-1, ISO 17636-S; or ISO 17640 |
| Surface Crack Detection by Penetrant Test or Magnetic Particle Test | 100 | ISO 3452-1 or ISO 17638 |

TABLE 2.10    Destructive testing

| Type of Test | Extend of Testing | Test Standard | Test results |
|---|---|---|---|
| Transverse Tensile Test | 2 Specimens | ISO 4136 | $f_y$ = 514 MPa, $f_u$ = 638 MPa |
| All-Weld Metal Test | 2 Specimens | ISO 4136 | $f_y$ = 483 MPa, $f_u$ = 592 MPa |
| Transverse Bend Test | 4 Specimens | ISO 5173 | No defect observed |
| Macroscopic Test | 1 Specimen | ISO 17639 | No defect observed |
| Charpy Impact Test (−20°C) | 2 Sets | ISO 9016 and ISO 1481 | 182–223 J |
| Hardness Test | Fusion zone, HAZ and parent metal | ISO 9015 | See Figure 2.8. |

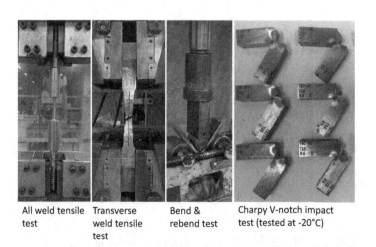

All weld tensile test    Transverse weld tensile test    Bend & rebend test    Charpy V-notch impact test (tested at -20°C)

FIGURE 2.7    Welding procedure specification (destructive) tests.

parts connected, provided that the weld is made with a suitable consumable which will produce all-weld tensile specimens having both a minimum yield strength and a minimum tensile strength not less than those specified for the parent metal. As can be seen in Table 2.9, both the transverse tensile specimens and the all-weld metal specimens showed adequate strengths in terms of yielding and tensile strengths. Specific attention was paid to the hardness distribution across the welding zone, which indicates the localized influence of welding heat input on the base material. As shown in Figure 2.8, the hardness in the heat-affected zone (HAZ) was comparable to the fusion zone but significantly lower than the base material. As mentioned earlier, TMCP steels obtain reply on the heat treatment process to refine their microstructure and achieve higher strength and hardness. Being sensitive to heat, the refined microstructure in the HAZ was inevitably affected by welding heat input in the localized area. Nevertheless, a minimum hardness of around 210 is guaranteed, which is equivalent to approximately 650–680 MPa of tensile strength.

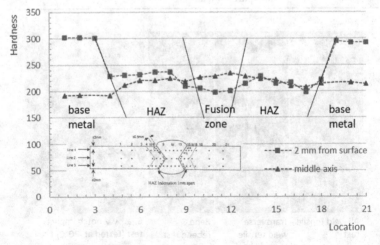

FIGURE 2.8   Hardness distribution across the welding zone.

## 2.3 REINFORCEMENT, PROFILED STEEL SHEETING AND SHEAR CONNECTORS

### 2.3.1 Reinforcement

EN 1994-1-1 refers the designers to EN 1992-1-1 for the properties of reinforcing steel. It is also recommended that for composite construction, EN 1994-1-1 has stated that the design value for elastic modulus of reinforcement steel may be taken as equal to that of structural steel in EN 1993-1-1, which is 210 GPa [22].

The yield strength and ductility of the ribbed weldable reinforcing steel material in either bar or fabric should be specified in accordance with the requirements of EN 10080:2005. The characteristic yield strength of reinforcement to these standards shall be between 400 MPa and 600 MPa, depending on the national market. To ensure sufficient ductility for plastic analysis, it is noted in EN 1992-1-1, Class B or Class C should be specified [23]. The reinforcement industry in Singapore commonly uses grade B500B reinforcing steel (complying with SS 560:2016 and/or SS 561:2010), which has the characteristic yield strength, $f_{yk}$, of 500 MPa. The design strength of reinforcement is given by EN 1992-1-1 as follows:

$$f_{yd} = \frac{f_{yk}}{\gamma_s} \qquad (2.6)$$

where

$f_{yk}$ is the yield strength (0.2% proof stress)

$\gamma_s$ is the partial factor for reinforcing steel, $\gamma_s = 1.15$

### 2.3.2 Profiled Steel Sheeting

The profile steel sheeting is generally manufactured in either a re-entrant or trapezoidal profile, as shown in Figure 2.9. The profile sheeting is formed using zinc-coated steel coil in grades S280 to S450. The height of the profile sheeting ($h_p$) is defined as the height of the shoulder of the profile, even if the profile of a re-entrant detail on the flange.

EN 1994-1-1 gives the minimum value of the nominal bare metal thickness as 0.7 mm [21]. In EN 1993-1-1, it is stated that a

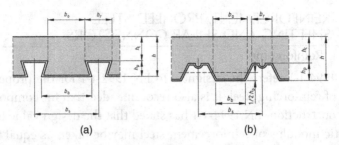

FIGURE 2.9 General profile steel sheeting. (a) Re-entrant trough profile; and (b) Open trough profile.

zinc coating of 275 g/m², total for both sides, is suitable for most non-aggressive interior conditions [24]. For exterior or aggressive interior conditions, additional protection is required [24]. If protective paint is specified, the method and timing of the application have to be considered, as painted steel sheeting is not suitable for the process of through-sheet welding of shear studs.

### 2.3.3 Shear Connectors

Stud connectors are extensively employed in steel-concrete composite structures to resist shear forces at the interface between the steel and concrete. In EN 1994-1-1, detailed provision is given to headed studs only. In general, the diameter of the headed stud varies from 10 to 25 mm and the length varies from 65 to 150 mm [22], while the 19 mm diameter and 100 mm length are most commonly used in actual projects. The design shear resistance of the shear connector can be determined from whichever is smaller:

For failure of stud:

$$P_{Rd} = \frac{0.8 f_u \pi d^2}{4 \gamma_v} = 0.16 f_u \pi d^2 \qquad (2.7)$$

where

$f_u$ is the ultimate tensile strength of the head stud but not more than 500 N/mm² for sheeting spanning parallel to supporting beam and not more than 450 N/mm² for sheeting spanning transversely to supporting beam

$d$ is the diameter of the shank of the headed stud

$\gamma_v$ is the partial factor shear stud, $\gamma_v = 1.25$

For failure of concrete:

$$P_{Rd} = \frac{0.29\alpha d^2 \sqrt{f_{ck} E_{cm}}}{\gamma_v} = 0.232\alpha d^2 \sqrt{f_{ck} E_{cm}} \qquad (2.8)$$

where

$d$ is the diameter of the shank of the headed stud

$f_{ck}$ is the characteristic cylinder strength of the concrete

$E_{cm}$ is the secant elastic modulus of the concrete (given in EN 1992-1-1)

$\alpha$ is the corrective factor

$\gamma_v$ is the partial factor shear stud, $\gamma_v = 1.25$

Corrective factor:

$$\alpha = 0.2\left(\frac{h_{sc}}{d} + 1\right) \text{when } 3 \le \frac{h_{sc}}{d} \le 4 \qquad (2.9)$$

$$\alpha = 1 \text{ when } \frac{h_{sc}}{d} > 4 \qquad (2.10)$$

Figure 2.10 illustrates the stress compatibility between 19 mm-headed shear studs and the concrete. It is demonstrated that the shear resistance will be governed by the concrete when the concrete grade is lesser than 45 MPa, and when the strength of concrete is higher than 45 MPa, the failure of shear stud can be expected.

## 2.3.4 Validation of Buildability

### 2.3.4.1 Experimental Investigation on the Resistance of the Shear Connection

To validate the compatibility among the selected HPGC, high-performance steel (HPS) S460M and the conventional headed shear studs, a modified push-out test adopted from EC4 was

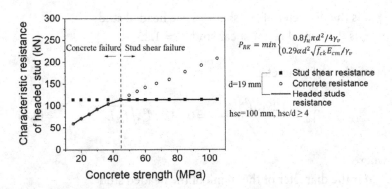

FIGURE 2.10 Shear resistance considering compatibility between shear stud and concrete strength.

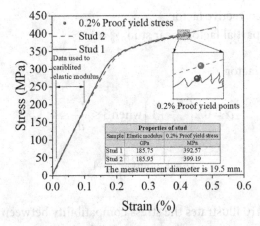

FIGURE 2.11 Tensile properties of shear stud.

carried out. Stress-strain curve of the 19 mm diameter-headed shear studs was obtained from quasistatic tensile test, as shown in Figure 2.11. Two specimens of modified push-out using HPGC design of S1 and S2 were planned and examined, with the schematic details in Figure 2.12.

The push-out specimens were planned with considerations of the mechanical properties of the headed shear studs, HPGC and the maximum compression capacity available in the Singapore Institute of Technology. Based on the stress compatibility (Figure 2.11) and

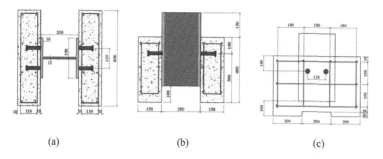

FIGURE 2.12 Schematic details of modified push-out test. (a) Top view (b) Side view; and (c) Front view.

FIGURE 2.13 Actual push-out test setup.

Equations 2.7 and 2.8, it is expected that the failure mode will be governed by the failure of stud, with a predicted shear resistance, $P_{Rd}$, of 113 kN. Two linear variable displacement transducers (LVDT) are installed to capture and record the displacement, with several strain gauges installed to monitor and record the stain history during the loading process, as shown in Figure 2.13.

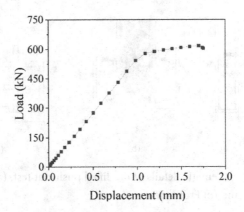

FIGURE 2.14    Load-displacement relationship.

The load-displacement curve of specimen S1, is presented in Figure 2.14 for example. The load-displacement in general exhibited a bi-linear relationship. It is observed that when the displacement approaches approximately 1 mm, linear curve changes the gradient as it approaches the designed peak load (452 kN). However, it is evident from the load-displacement curve that the shear resistance of the specimen was beyond the design load. This phenomenon is likely to be caused by the over-reinforced welding technique used by the manufacturer—as a ceramic ring, 3 mm in height and 21 mm diameter, was not deployed for the welding process. This ring is usually used for quality control and assurance of the welding of the shear stud to steel sections. Therefore, the finite element (FE) analysis will be used for verification of the shear resistance, with due consideration given to the ceramic ring. Nonetheless, the push-out test conducted has successfully verified the material compatibility, between for HPS and HPGC.

Table 2.11 compares the results of the two specimens, namely S1 and S2. Noted that the compression test machine was not able to completely fail the specimens during testing due to limited capacity. The load-displacement curve (Figure 2.14) does not

TABLE 2.11   Test results comparison between
S1 and S2 Specimen

|  | S1 | S2 |
|---|---|---|
| Design strength (kN) | 454 | 454 |
| Yield load (kN) | 581 (+28%) | 579 (+27.5%) |

include the post-peak behaviour. Based on the load-displacement relationship curves obtained, the equivalent yield can be determined using the equivalent yield method. The equivalent yield for S1 and S2 were 581 and 579 kN, respectively. This shear resistance is 27% higher than the designed shear resistance predicted by Equations 2.8 and 2.9.

### 2.3.4.2 Finite Element Analysis

The FE model is constructed based on a quarter of the actual specimen taking advantage of the symmetry, as shown in Figure 2.15a; the completed FE model consisting of all connection components used, namely HPGC slab, the HPS section, 19 mm diameter-headed shear studs, welding ring and reinforcing steel bars, is as shown in Figure 2.15b.

The general contact interaction procedure with normal behaviour ("hard" formulation) and tangential behaviour ("penalty"

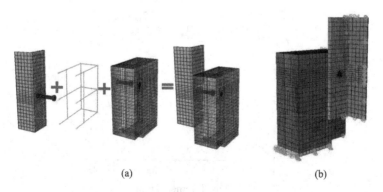

(a)                                    (b)

FIGURE 2.15   Details of the FE model. (a) Model detail and (b) Boundaries.

friction formulation) was used. The friction coefficient of 0.4 was set for shear studs, and the surface between the HPS and HPGC. Reinforcement bars were modelled to be embedded in the concrete slab. The adjacent surfaces of the specific concrete component were completely fixed to the "Support" reference point, as depicted in Figure 2.16b. Subsequently, the reference point was assigned a fully fixed boundary condition, except for lateral translation U3, which represents motion in the Z-direction. Loading up to failure was applied as a vertical displacement U2 = 6 mm of the "Jack" reference point to which the top surface of the steel section was constrained; refer to Figure 2.15b.

Hexahedron finite elements (C3D8) were used for model parts (concrete slab, shear studs and I beam section bolts). Mesh size varied for different parts in terms of their size and importance. For example, stud shaft was meshed and 8.7 mm (in height direction) elements, while the ring had a mesh size of 1.5 mm (in height direction). A global mesh of 20 mm was chosen for concrete slab and reinforcement bar as well as the I beam section.

### 2.3.4.3 Comparison between Test and Numerical Analysis Results

The load-displaceme. nt curve obtained from the FE model analysis is compared to the actual load-displacement curve, as presented in Figure 2.16. The load-displacement from FE model and

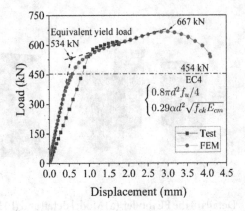

FIGURE 2.16  Load-displacement curves of testing and FE analysis.

actual is good agreement. The bearing capacity from the FE model was approximately 667 kN, with a corresponding displacement of 3 mm. This shear resistance is 46.9% higher than the required design shear resistance of 454 kN determined by using the EC4 Equations (2.7) and (2.8).

As numerical analyses performed on FE models showed good agreement with experimental results, the equivalent yield method [25] is used to determine the shear resistance. The procedure for equivalent yield is as follows:

1) A linear fitting in the initial stage and plastic stage is performed.

2) Extrapolate the two linear fitting curves until they intersect with each other.

3) The intersection point of the two fitting curves is the equivalent yield load.

## 2.4 SUMMARY

Several approaches for creating high-performance green concrete C55/67, including the use of RCA, GGBS and silica fume, are evaluated. Recommendation on carbonation of RCAs is also given to enhance the properties of RCAs, which are proven to be the weak link in high-performance concrete. The second focus of this chapter is on the WPS and the relevant tests for quality control during fabrication of the pre-engineered sections, as the TMCP steel in grade S460M is predominantly available in plate forms. Lastly, to verify the buildability of HPCB, shear connection specimens were fabricated by using the selected HPGC design mixes, S460M built-up sections qualified by the WPS and standard-headed shear studs. Through the experimental studies, the following conclusions are drawn:

1. The use of 10–20% carbonated RCA to replace natural coarse aggregates is shown to have noticeable beneficial effects (2–15% improvement) on to compressive strength of

high-performance concrete C55/67 when pure OPC is used as a binder. However, when 20% GGBS and 3.5% silica fume are used to partially replace OPC, the synergy between this blended binder and carbonated RCA tends to adversely affect the development of strength by approximately 10%.

2. The selected HPGC, high-performance steel S460M and the conventional headed shear studs are shown to be compatible through the buildability test. The pull-out strengths of the shear connection specimens were shown to be 27% higher than the design shear resistance predicted by the EC4 equations. This result was further supported by the finite element analysis.

## REFERENCES

1. Gagg, C. R. Cement and concrete as an engineering material: An historic appraisal and case study analysis. *Engineering Failure Analysis* 40, 114–140 (2014).
2. Cement industry accounts for about 8% of $CO_2$ emissions. One startup seeks to change that – CBS News (2023).
3. Cement carbon dioxide emissions quietly double in 20 years – CAN (2022).
4. Fundamentals of High-Performance Concrete - Edward G. Nawy (2001).
5. Jones, C. Inventory of Carbon & Energy Database V3.0, Circular Ecology, 2019.
6. Nielsen, C. V. & Glavind, M. Danish experiences with a decade of green concrete. *Journal of Advanced Concrete Technology* 5, 3–12 (2007).
7. Habert, G., Arribe, D., Dehove, T., Espinasse, L. & Roy, R. Le. Reducing environmental impact by increasing the strength of concrete: Quantification of the improvement to concrete bridges. *Journal of Cleaner Production* 35, 250–262 (2012).
8. Isaia, G. C. High-performance concrete for sustainable constructions. *Waste Management Series* 1, 344–354 (2000).
9. Sivakrishna, A., Adesina, A., Awoyera, P. O. & Kumar, K. R. Green concrete: A review of recent developments. *Mater Today: Proceedings* 27, 54–58 (2020).

10. Tae, S., Baek, C. & Shin, S. Life cycle $CO_2$ evaluation on reinforced concrete structures with high-strength concrete. *Environmental Impact Assessment Review* 31, 253–260 (2011).
11. Detwiler, R. J. & Kumar Mehta, P. Chemical and physical effects of silica fume on the mechanical behavior of concrete. *ACI Materials Journal* 86, 609–614 (1989).
12. NUS-ESI, *Building Embodied Carbon Calculator (BECC)* (2022).
13. Yang, K. H., Jung, Y. B., Cho, M. S., & Tae, S. H. Effect of supplementary cementitious materials on reduction of $CO_2$ emissions from concrete. *Journal of Cleaner Production* 103, 774–783 (2015).
14. Juenger, M. C. G. & Siddique, R. Recent advances in understanding the role of supplementary cementitious materials in concrete. *Cement and Concrete Research* 78, 71–80 (2015).
15. Pu, Y., et al. Accelerated carbonation technology for enhanced treatment of recycled concrete aggregates: A state-of-the-art review. *Construction and Building Materials* 282, 122671 (2021).
16. Daneti, S. B. & Tam, C. T. Sustainability of concrete constructions: The role of materials and practices. *Lecture Notes in Civil Engineering* 61, 381–395 (2020).
17. Liang, C., Lu, N., Ma, H., Ma, Z. & Duan, Z. Carbonation behavior of recycled concrete with $CO_2$-curing recycled aggregate under various environments. *Journal of $CO_2$ Utilization* 39, 101185 (2020).
18. Zhan, B., Poon, C. S., Liu, Q., Kou, S., & Shi, C. Experimental study on $CO_2$ curing for enhancement of recycled aggregate properties. *Construction and Building Materials* 67, 3–7 (2014).
19. Fu, Q., Zhang, Z., Zhao, X., Xu, W. & Niu, D. Effect of nano calcium carbonate on hydration characteristics and microstructure of cement-based materials: A review. *Journal of Building Engineering* 50, 104220.(2022).
20. Awoyera, P. O., Perumal, P., Ohenoja, K. & Mansouri, I. Upcycling $CO_2$ for enhanced performance of recycled aggregate concrete and modeling of properties. In *The Structural Integrity of Recycled Aggregate Concrete Produced With Fillers and Pozzolans*, pp. 349–364. (2022).
21. Li, L., Liu, Q., Huang, T. & Peng, W. Mineralization and utilization of $CO_2$ in construction and demolition wastes recycling for building materials: A systematic review of recycled concrete aggregate and recycled hardened cement powder. *Separation and Purification Technology* 298, 121512 (2022).
22. EN 1994-1-1 Eurocode 4: Design of Composite Steel and Concrete Structures- Part 1-1: General rules and rules for buildings (2005).

23. EN 1992-1-1 Eurocode 2: Design of concrete structures - Part 1-1: General rules and rules for buildings (2005).
24. EN 1993-1-1 Eurocode 3: Design of steel structures - Part 1-1: General rules and rules for buildings (2005).
25. Al-Khatab, Z. & Bouchaïr, A. Analysis of a bolted T-stub strengthened by backing-plates with regard to Eurocode 3. *Journal of Constructional Steel Research* 63, 1603–1615 (2007).
26. Feng, P., Chang, H., Xu, G., Liu, Q., Jin, Z. & Liu, J. Feasibility of utilizing recycled aggregate concrete for revetment construction of the lower Yellow River. *Materials* 12, 4237 (2019).

# CHAPTER 3

# HPCB at Ultimate Limit State

## 3.1 ULTIMATE LIMIT STATE (ULS)

### 3.1.1 Action

The actions that are to be taken into consideration in verification of the beams are permanent and variable actions. The former consists of the self-weight of steel sections, composite slab, finishes and services. The latter is the live loads made up of the allowance for occupancy loads and movable partitions.

### 3.1.2 Combination of Actions

The combination of actions is stipulated in EN 1990 (EC0) (Equations 3.1, 3.2 and 3.3). Considering the three equations, the most onerous case is given by (3.10). Based on this fundamental combination of actions and the partial factor values stated in the Singapore national annexes, the combination of action is to be considered for high-performance pre-engineered steel concrete composite beam (HPCB) during the final composite stage.

DOI: 10.1201/9781032626932-3

$$\sum_{j\geq1}\gamma_{G,j}G_{k,j}+\gamma_pP+\gamma_{Q,1}Q_{k,1}+\sum_{j\geq1}\gamma_{Q,i}\Psi_{0,1}Q_{k,i} \qquad (3.1)$$

$$\sum_{j\geq1}\gamma_{G,j}G_{k,j}+\gamma_pP+\gamma_{Q,1}\Psi_{0,1}Q_{k,1}+\sum_{j\geq1}\gamma_{Q,i}\Psi_{0,1}Q_{k,i} \qquad (3.2)$$

$$\sum_{j\geq1}\xi_j\gamma_{G,j}G_{k,j}+\gamma_pP+\gamma_{Q,1}Q_{k,1}+\sum_{j\geq1}\gamma_{Q,i}\Psi_{0,1}Q_{k,i} \qquad (3.3)$$

## 3.2 PLASTIC RESISTANCE WITH FULL SHEAR CONNECTION

### 3.2.1 Steel Concrete Strain Compatibility

Figure 3.1 illustrates the typical development of a composite beam's triangular bending stress distribution when the maximum shear capacity is attained. Based on the position of the plastic neutral axis (PNA), a simplified linear relationship can be established between the strain of the extreme concrete fibre and the strain of the most extreme steel fibre. In an ideal situation, under the ULS

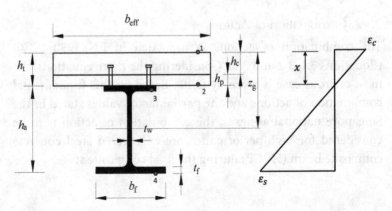

FIGURE 3.1 Schematic strain state of the HPCB subjected to bending moment.

and strain compatible condition, concrete and steel yield simultaneously and result in a rectangular stress block.

In order to ensure the strain compatibility of composite structures for a cost-effective design, EN 1994-1-1 restricts the nominal yield strength of materials for composite structural design. Firstly, the grade of concrete is limited to C60/75, followed by 600 N/mm² for reinforcement steel bars, and, lastly, 460 N/mm² for structural steel sections [1]. Figure 3.2 illustrates the theoretical yielding strain of various structural steel and the failure strain of HPGC in grades C55/67 and C60/75 in relation to EC3 and EC2, respectively, in order to ascertain the strain compatibility of the proposed HPCB. In Figure 3.2, it is clearly demonstrated that S460 structural steel (yield strain of 0.0022) is strain compatible with concrete grade up to C60/75 (failure strain of 0.0029).

A simplified mathematical model can be developed to verify the strain compatibility between the two materials by using the similar triangle concept. The position of PNA is measured from the extreme fibre of concrete, "$x$", and the strain at PNA is "0" (Figure 3.1). The failure strain of concrete is $\varepsilon_c$. The yield strain of steel is $\varepsilon_s$. Therefore, using the similar triangle concept, the strain

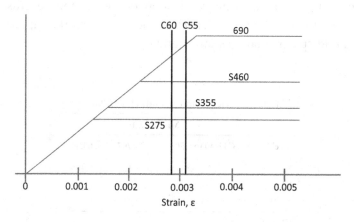

FIGURE 3.2 Strain compatibility between the structural steel and concrete.

of the extreme fibre of the steel can be determined from the following equation:

$$\varepsilon_{sy} = \frac{(h-x) \times \varepsilon_{cf}}{x} \quad (3.4)$$

where
  $\varepsilon_{sy}$ is the yield strain of structural steel
  $\varepsilon_{cf}$ is the failure strain of concrete
  $h$ is the total height of composite beam
  $x$ is the distance of PNA from extreme concrete fibre

Based on Equation (3.4), the minimum height ratio can be determined, as shown in Table 3.1.

## 3.2.2 Effective Section

The initial stage in the determination of a composite cross-section involves the evaluation of the width of concrete flange that will effectively contribute to the composite behaviour alongside the steel section. The effective width is expressed in relation to the span of the beam, and different values apply at different points along the beam. For the verification of the composite cross-section, the distribution of effective width between the supports and mid-span region is as shown in Figure 3.3.

TABLE 3.1    Height to PNA ratio for HPCB

|  | Min $h/x$ ratio | | |
| --- | --- | --- | --- |
| Steel Grade | C32/40 to C50/60 | C55/67 | C60/75 |
| S275 | 1.37 | 1.42 | 1.45 |
| S355 | 1.49 | 1.55 | 1.59 |
| S460 | 1.63 | 1.71 | 1.76 |

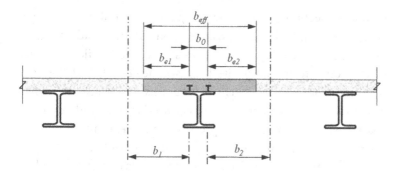

FIGURE 3.3   Effective concrete flange width.

The effective width at mid-span may be taken as:

$$b_{\text{eff}} = b_0 \sum b_{\text{ei}} \tag{3.5}$$

where

   $b_0$ is the distance between the centres of the outstand shear connectors

   $b_{\text{ei}}$ is the value of the effective width of the concrete flange on each side of the web, taken as $L_e/8$ but not greater than $b_{\text{ei}}$

   $L_e$ is length between points of zero bending moment

The effective width for the last quarter of span to support can be assumed to reduce and can be determined by:

$$b_{\text{eff}} = b_0 \sum \beta_i b_{\text{ei}} \tag{3.6}$$

where

$$\beta_i = \left(0.55 + 0.025 L_e / b_{\text{ei}}\right) \le 1.0$$

## 3.2.3  Rigid Plastic Theory

In the absence of pre-stressing due to tendons, rigid plastic theory can be used to determine the bending resistance for Class 1 and 2 composite sections, and elastic analysis or the nonlinear theory is acceptable for all other composite sections [1]. In consideration for designing the bending moment resistance, the tensile resistance of the concrete in composite beams is neglected, and the contribution from profiled steel sheeting must be ignored too when in compression. The composite cross-section is considered to be Class 1 when concrete is compression and steel beam in tension. Therefore, the bending resistance of the composite beam is typically taken as plastic bending resistance, while the elastic bending resistance is not considered [1].

Three typical stress block distribution scenarios for composite beams with full shear connection are shown in Figure 3.4 [6]. Concrete in compression is assumed to resist a stress equal to 0.85 $f_{cd}$ over the full depth from the extreme compressed fibre to the PNA [1]. The amount of concrete available for resisting the compressive force is limited by the effective width ($b_{eff}$) of the concrete flange and the depth of concrete to the profiled steel sheeting. The concrete in the profiled steel sheeting is normally ignored for both primary and secondary beams. The contribution of the reinforcement and steel sheeting is generally ignored as the bending resistance is small.

## 3.2.4 PLASTIC MOMENT RESISTANCE

The design of plastic resistance adopts the same approach as Eurocode 4. However, to optimize the efficiency of the composite beam, the PNA shall be maintained within the concrete slab to create an efficient solution, as shown in (3.5).

The bending resistance for HPCB can be obtained through the following equations.

Compressive resistance of concrete:

$$N_{c,f} = h_c b_{eff} 0.85 f_{cd} \qquad (3.7)$$

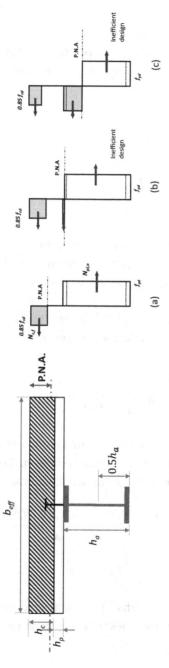

FIGURE 3.4 Typical PNA location in composite beams.

where

$h_c$ is the depth of the concrete slab

$b_{eff}$ is the effective flange width of concrete slab

$f_{cd}$ is the design value of compressive strength of concrete

$$\left(\frac{f_{ck}}{\gamma_c}\right) = \frac{f_{ck}}{1.5}$$

Tensile resistance of structural steel section:

$$N_{pl.a} = \frac{A_a f_y}{\gamma_a} \tag{3.8}$$

where

$A_a$ is the area of the structural steel section

$f_y$ is the design yield strength of structural steel

$\gamma_a$ is the appropriate partial factor for structural steel, $\gamma_a = 1.0$

Note that in an efficient HPCB design, usually $N_{pl.\,a} < N_{c,f}$ such that the PNA will remain in the concrete slab.

Location of PNA:

$$x = \frac{N_{pl,a}}{b_{eff} \, 0.85 f_{cd}} < h_c \tag{3.9}$$

As the pre-engineered section is an asymmetrical I-shaped, the centre of gravity will shift closer to the bottom flange (Figure 3.5). It can be determined through the second moment of area:

$$h_s = \frac{\sum (A_i y_i)}{A_a} = \frac{b_{f1} t_{f1} \left( h_a - 0.5 t_{f1} \right) + h_w t_w \left( 0.5 h_w + t_{f2} \right) + 0.5 b_{f2} t_{f2}^2}{A_a} \tag{3.10}$$

where

$b_{f1}$ is the top flange width of pre-engineered steel section

$t_{f1}$ is the top flange thickness of pre-engineered steel section

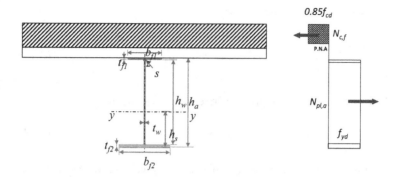

FIGURE 3.5   Ideal composite action (PNA in slab).

$h_w$ is the height of web of pre-engineered steel section
$t_w$ is the thickness of web of pre-engineered steel section
$b_{f2}$ is the bottom flange width of pre-engineered steel section
$t_{f2}$ is the bottom flange thickness of pre-engineered steel section

Determine the moment resistance of composite section:

$$M_{pl,Rd} = N_{pl,a}\left[ h_a - h_s + h_c + h_p - \left(\left(\frac{N_{pl.a}}{N_{c,f}}\right) \times \left(\frac{h_c}{2}\right)\right)\right] \quad (3.11)$$

Or

$$M_{pl,Rd} = N_{pl,a}\left[ h_a - h_s + h_c + h_p - 0.5x \right] \quad (3.12)$$

## 3.3  VERTICAL SHEAR RESISTANCE

It is assumed that the vertical shear due to the factored loading is resisted by the steel section only. The shear resistance should be in reference to EN 1993-1-1 clause 6.2.6 [2].

The shear resistance:

$$V_{pl,a,Rd} = A_v \frac{f_{yd}}{\sqrt{3}} \qquad (3.13)$$

The shear area of the pre-engineered steel section:

$$A_v = h_w t_w \qquad (3.14)$$

The shear buckling of an unstiffened steel web can be neglected if:

$$\frac{h_w}{t_w} \leq 72\varepsilon \qquad (3.15)$$

where

$$\varepsilon = \sqrt{\frac{235}{f_y}}$$

If a slender web is used, it is suggested to adopt the buckling check stated in EN1993-1-5 [3]. However, a conservative web design satisfying Eqn. (3.15) is recommended.

## 3.4 SHEAR CONNECTION

The design rules for determining the resistance of headed studs as shear connectors with profiled steel sheeting are given in EN 1994-1-1. The design resistance of a head stud shear connector in solid concrete slab is the smaller of failure of stud or failure of concrete. The rules for profile steel sheeting comprise sheeting spanning parallel to the supporting beam and sheeting transverse to the supporting beam. Therefore, the resistance of a headed stud within the profile steel sheeting (Figure 3.6) is determined by multiplying the design resistance of the head stud connector in solid concrete slab by a reduction factor, $k_l$ for parallel and $k_t$ transverse.

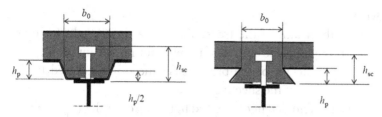

FIGURE 3.6  Shear stud in profile sheeting.

Design resistance for failure of stud:

$$P_{Rd} = \frac{0.8 f_u \pi d^2}{4 \gamma_v} = 0.16 f_u \pi d^2 \tag{3.16}$$

where
  $f_u$ is the ultimate tensile strength of the head stud but not more
    than 500 N/mm² for sheeting spanning parallel to support-
    ing beam and not more than 450 N/mm² for sheeting span-
    ning transversely to supporting beam
  $d$ is the diameter of the shank of the headed stud

Design resistance for failure of concrete:

$$P_{Rd} = \frac{0.29 \alpha d^2 \sqrt{f_{ck} E_{cm}}}{\gamma_v} = 0.232 \alpha d^2 \sqrt{f_{ck} E_{cm}} \tag{3.17}$$

where
  $f_{ck}$ is the characteristic cylinder strength of the concrete
  $E_{cm}$ is the secant elastic modulus of the concrete (given in EN
    1992-1-1)

The reduction factor, $k_l$, can be determined by:

$$k_l = 0.6 \frac{b_0}{h_p} \left( \frac{h_{sc}}{h_p} - 1 \right) \leq 1.0 \tag{3.18}$$

where

$b_0$ is the width of the trapezoidal rib at mid-height for open through profile and the minimum for re-entrant profile

$h_p$ is the height of the profile steel sheeting measured to the shoulder of the sheeting

$h_s$ is the weld height of the stud but not greater than $h_p + 75$ mm

The reduction factor, $k_t$, can be determined by:

$$k_t = \frac{0.7}{\sqrt{n_r}} \frac{b_0}{h_p} \left( \frac{h_{sc}}{h_p} - 1 \right) \le k_{t,\max} \tag{3.19}$$

where

$n_r$ is the number of stud connectors in one rib, not exceeding 2

## 3.5 YIELD LINE APPROACH

### 3.5.1 Material Model

Concrete is classified on the basis of its compressive strength, and the grade of the concrete is based on the corresponding specific value of its characteristic strength, $f_{ck}$ [4]. The use of the characteristic compressive strength $f_{ck}$ is utilized in the study and design of concrete structures. The determination of this value can be obtained by a strength assessment employing the condition that the value $f_{ck}$ should not be exceeded by more than 5% of all feasible strength measurements for the specified concrete [4]. The elastic modulus of normal density concrete with natural particles may be approximated by using the specific strength. In accordance with EN 1992-1-1, the design of cross sections can employ three stress–strain relation models, namely the parabola-rectangle, the bilinear and the rectangle stress distribution models, as seen in Figure 3.7. The yield line approach analytical model in EN 1992-1-1 specifies the use of the simplified bilinear stress–strain relation, which is deemed to be equal or more conservative than the

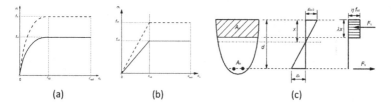

FIGURE 3.7 Concrete stress-strain curve models (EN 1992-1-1). (a) Parabola-rectangle diagram for concrete under compression; (b) Bilinear stress-strain relation; and (c) Rectangular stress distribution in cross-section.

parabola-rectangle [5]. Another reason for the adoption of the bilinear stress–strain model is the simplicity of the mathematical equations used to create the stress-block and strain relationship for compression and tension forces. The model's design value is selected as the design strength, denoted as $f_{cd} = f_{ck}/\Upsilon$, in accordance with the recommendations provided in EN 1992-1-1 and EN 1994-1-1.

The stress–strain relation for hot-rolled structural quasistatic tensile load is illustrated in Figure 3.8 [6]. 210 GPa is taken as the recommended elastic modulus during the elastic stage [3]. As the yield stress, $f_y$, reaches the corresponding strain of $\varepsilon_y$, the strain continues to increase along a yield plateau without any increase in stress. As the strain continues to increase to the point of strain hardening, strain hardening initiates and the stress will start to increase again with the increasing strain. Ultimately, the stress attains its greatest magnitude at the point of ultimate tensile strength, $f_u$, accompanied by the equivalent strain, $\varepsilon_u$. Once the strain and tension are beyond the ultimate threshold, the phenomenon of necking starts, leading to the final fracture. According to the EN 1993-1-1 standard, it is advised to employ the bilinear stress–strain correlation for the structural steel grades listed in Section 3.3 of EN 1993-1-1 for conducting plastic global analysis [3]. The selected model implies that the structural steel does not

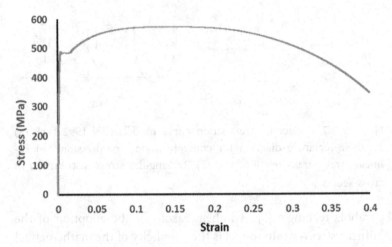

FIGURE 3.8  Stress–strain of S460 plates.

undergo strain hardening. Consequently, an elastic-fully plastic model is employed, with a yield strength ($f_y$) of 460 N/mm² and a yield strain ($\varepsilon_y$) of 0.00219.

### 3.5.2 Strain–Stress Development

The yield line analysis, originally developed by Johansen during the 1940s, is a well-established method used for predicting the maximum load that can be sustained. This analysis involves identifying critical yield lines that developed under increasing load. The yield line theory is based on many assumptions: firstly, the plane section of the structure remains in the same plane; secondly, the yield lines are straight; thirdly, the yield lines terminate at the boundaries of the structure; and, fourthly, the yield line must cross the axis of the structure. Based on the assumption and the principle of yield line approach, three analysis models, for different stages (elastic, elastic-plastic and fully plastic) can be developed, as shown in Figure 3.9.

The three models consider the varying positions of the neutral axis. Initially, it is important to note that the neutral axis is located within the pre-engineered steel section. At this particular time, it

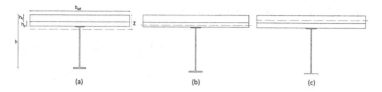

FIGURE 3.9 Model for the yield model. (a) Neutral axis in steel; (b) Neutral axis in profile steel sheeting; and (c) Neutral axis in concrete.

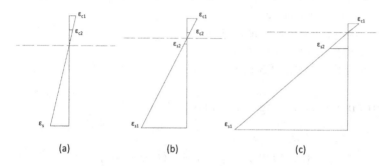

FIGURE 3.10 Strain development. (a) Linear elastic stage; (b) Yielding stage; and (c) Plastic stage (large deformation).

is appropriate to regard the section as being in the elastic phase. Additionally, the neutral point is situated within the profiled steel sheets. At this juncture, a portion of the steel section has already undergone yielding, while the composite section is now experiencing partial plastic deformation. Finally, the neutral axis is situated within the concrete slab, where the majority of the pre-engineered part has experienced yielding, and the composite section is said to be in its plastic phase. The determination of the yield line for each model, as seen in Figure 3.10, may be achieved by considering the position of the neutral axis. By using the notion of comparable triangles, as explained in Section 3.2, it is possible to establish the relationship between the strain of steel and concrete.

Strain relation during elastic stage:

$$\frac{\varepsilon_s}{(h-z)} = \frac{\varepsilon_{c1}}{x} = \frac{\varepsilon_{c2}}{(h_c - z)} \qquad (3.20)$$

where

$\varepsilon_s$ is the strain of extreme steel fibre

$\varepsilon_{c1}$ is the strain of extreme concrete fibre

$\varepsilon_{c2}$ is the strain of concrete fibre at intersection of concrete and profile sheeting

$z$ is the location of neutral axis

$h$ is the height of HPCB

$h_c$ is the height of concrete

Strain relation during elastic-plastic stage:

$$\frac{\varepsilon_{s1}}{(h-z)} = \frac{\varepsilon_{c1}}{z} = \frac{\varepsilon_{c2}}{(x-h_c)} = \frac{\varepsilon_{s2}}{(h_c + h_p - z)} \qquad (3.21)$$

where

$\varepsilon_{c1}$ is the strain of extreme concrete fibre

$\varepsilon_{c2}$ is the strain of concrete fibre at intersection of concrete and profile sheeting

$\varepsilon_{s1}$ is the strain of extreme steel fibre

$\varepsilon_{s2}$ is the strain of steel web

$h_p$ is the height of profile sheeting

Strain relation during the plastic stage:

$$\frac{\varepsilon_{s1}}{(h-z)} = \frac{\varepsilon_c}{z} = \frac{\varepsilon_{s2}}{(h_c + h_p - z)} \qquad (3.22)$$

$\varepsilon_c$ is the strain of extreme concrete fibre

$\varepsilon_{s1}$ is the strain of extreme steel fibre

$\varepsilon_{s2}$ is the strain of steel web

$z$ is the location of neutral axis

Based on the elastic-perfectly plastic model, the stress blocks of each analysis model can be developed from the yield line, as shown in Figure 3.11. It shall be mentioned that the contribution of the top flange is ignored in this analysis as the contribution, lesser than 5%, to moment resistance is insignificant as compared to the bottom flange and web. The area of the stress block can be determined through equations from the stress–strain relation.

Stress block from stress–strain relation in elastic stage:

$$F_c = \frac{1}{2}\left(\sigma_{c1} + \sigma_{c2}\right) \cdot h_c \cdot c_f \tag{3.23}$$

$$F_t = \sigma_s \cdot t_{f1} \cdot b_{f1} + \frac{1}{2}\sigma_s \cdot \left(h - t_f - z\right) \cdot t_w \tag{3.24}$$

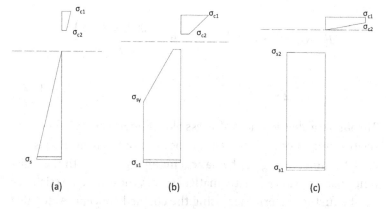

FIGURE 3.11  Stress block development. (a) Linear elastic stage; (b) Yielding stage; and (c) Plastic stage (large deformation).

where

$\sigma_{c1}$ is the compressive stress ($\varepsilon_{c1}E_{cm}$)

$\sigma_{c2}$ is the compressive stress ($\varepsilon_{c2}E_{cm}$)

$E_{cm}$ is the secant modulus of concrete

$\sigma_s$ is the tensile stress ($\varepsilon_s E_s$)

$E_s$ is the elastic modulus of steel

Stress block from stress–strain relation in elastic-plastic stage:

$$F_c = \frac{1}{2}\left(\sigma_{c1}+\sigma_{c2}\right) \cdot h_c \cdot c_f \tag{3.25}$$

$$F_t = \sigma_{s1} \cdot t_f \cdot b_f + \frac{1}{2}\left(\sigma_{sy}\right)\left( h - \frac{0.00219\left(h-z\right)}{\varepsilon_{s1}} \right)$$

$$+ \frac{1}{2}\left(\sigma_{sy}+\sigma_{s2}\right) \cdot t_w \cdot h_w \tag{3.26}$$

Stress block from stress–strain relation in plastic stage:

$$F_c = \frac{1}{2}\left( x + \left( x - \frac{\grave{o}_{c2}}{\grave{o}_{c1}} \right) \right) \cdot \sigma_{c1} \cdot c_f \tag{3.27}$$

$$F_t = \sigma_{s1} \cdot t_{f1} \cdot b_{f1} + \frac{1}{2}\left(\sigma_{sy}\right)\left( h - \frac{0.00219\left(h-x\right)}{\varepsilon_{s1}} \right)$$

$$+ \frac{1}{2}\left(\sigma_{sy}+\sigma_{s2}\right) \cdot t_w \cdot h_w \tag{3.28}$$

Through the development of stress blocks, together with the strain relationship at critical points, the location of the neutral axis can be determined using mathematical manipulation. With the assistance of an advanced mathematical solver, the moment resistance can be further determined using the coupled concept. A detailed working example for designing the HPCB under the ULS can be found in Appendix A.

## LIST OF SYMBOLS

$A_a$  is the area of the structural steel section

$A_v$  is the shear area of the pre-engineered steel section

$b_0$  is the distance between the centres of the outstand shear connectors/the width of the trapezoidal rib at mid-height for open through profile and the minimum for re-entrant profile

$b_{eff}$  is the effective flange width of concrete slab

$b_{ei}$  is the value of the effective width of the concrete flange on each side of the web, taken as $L_e/8$ but not greater than $b_{ei}$

$b_{f1}$  is the width of steel flange

$b_{f2}$  is the bottom flange width of pre-engineered steel section

$d$  is the diameter of the shank of the headed stud

$E_{cm}$  is the secant modulus of concrete

$E_s$  is the elastic modulus of steel

$f_{cd}$  is the design value of compressive strength of concrete

$f_{ck}$  is the characteristic cylinder strength of the concrete

$f_y$  is the design yield strength of structural steel

$f_u$  is the ultimate tensile strength of the head stud

$h$  is the total height of the composite beam

$h_c$  is the depth of the concrete slab

$h_p$  is the height of the profile steel sheeting

$h_s$  is the weld height of the stud

$h_w$  is the height of web of pre-engineered steel section

$L_e$  is length between points of zero bending moment

$n_r$  is the number of stud connectors in one rib, not exceeding 2

$t_{f1}$  is the top flange thickness of pre-engineered steel section

$t_{f2}$  is the bottom flange thickness of pre-engineered steel section

$t_w$  is the thickness of web of pre-engineered steel section

$x$  is the distance of PNA from extreme concrete fibre

$z$  is the location of neutral axis

$\varepsilon_c$  is the strain of concrete

$\varepsilon_{cf}$  is the failure strain of concrete

$\varepsilon_s$  is the strain of structural steel

$\varepsilon_{sy}$  is the yield strain of structural steel

$\gamma_a$  is the appropriate partial factor for structural steel

$\sigma_c$  is the concrete compressive stress

$\sigma_s$  is the steel tensile stress

$\sigma_{sy}$  is the steel tensile yield stress

## REFERENCES

1. BSI. BS EN 1994-1-1. Eurocode 4: Design of composite steel and concrete structures – Part 1-1: General rules and rules for buildings, British Standards Institution (2005).
2. BSI. BS EN 1993-1-1 Eurocode 3: Design of steel structures – Part 1-1: General rules and rules for buildings, British Standards Institution (2005).
3. BSI. BS EN 1993-1-5 Eurocode 3: Design of steel structures – Part 1-5: Plated structural elements, British Standards Institution (2003).
4. Walraven, JC., & van der Horst, AQC. FIB model code for concrete structures 2010. Internation Federation for Structural Concrete (fib) (2010).
5. BSI, BS EN 1992-1-1 Eurocode 2: Design of concrete structures – Part 1-1: General rules and rules for buildings, British Standards Institution (2005).
6. Gardner, L., Yun, X., Fieber, A. & Macorini, L. Steel design by advanced analysis: Material modelling and strain limits. *Engineering* 5, 243–249 (2019).

# CHAPTER 4

# HPCB at Serviceability Limit State

## 4.1 SERVICEABILITY LIMIT STATE

### 4.1.1 Action

The various actions taken into consideration for the serviceability limit state (SLS) include the various imposed load, forces and environmental factors that a structure must withstand while still maintaining its functionality and usability [1]. The typical considerations of SLS include deflection, vibration, cracking, serviceability of non-structural elements, comfort and durability.

### 4.1.2 Combination of Actions

The combination of actions is stipulated in EN 1990 [2], Equations (4.1), (4.2) and (4.3), namely characteristic combination, frequent combination and quasi-permanent combination, respectively. The choice of combination depends on the effect being considered.

The characteristic combination is an irreversible limit state, which includes impairment of functional performance and damage to structural elements, non-structural elements and finishes. The equation for characteristic combination is as follows:

DOI: 10.1201/9781032626932-4

$$\sum_{j\geq1}G_{k,j}+P+Q_{k,1}+\sum_{j\geq1}\Psi_{0,1}Q_{k,i} \qquad (4.1)$$

The frequent combination is a reversible limit state that will influence the comfort of the users, such as dynamic effects. The equation for frequent combination is as follows:

$$\sum_{j\geq1}G_{k,j}+\gamma_pP+\Psi_{1,1}Q_{k,1}+\sum_{j\geq1}\Psi_{2,1}Q_{k,i} \qquad (4.2)$$

The quasi-permanent combination takes into consideration long-term effects such as creep and cases, where deflections are only likely to influence the appearance of structure. The equation for quasi-permanent combination is as follows:

$$\sum_{j\geq1}G_{k,j}+P+\sum_{j\geq1}\Psi_{2,1}G_{k,i} \qquad (4.3)$$

## 4.2 GENERAL CRITERIA

Figure 4.1 shows the stress state of the HPCB subjected to bending moment. The moment of inertia is a pivotal feature in the majority of the SLS criterion. Nevertheless, the task of calculating the corresponding second moment of area becomes more difficult when building HPCB that incorporates asymmetrical pre-engineered steel sections with diverse geometric characteristics. This section examines the procedures for calculating the equivalent second moment of area and explores several deflection models for various loading scenarios. The determination of the equivalent second moment of area may be derived from the following equations:

Position of elastic neutral axis from the top fibre of the concrete slab

$$A_a\left(h_t+h_a-h_c-h_s\right)>\frac{b_{\text{eff}}h_c^2}{2n} \qquad (4.4)$$

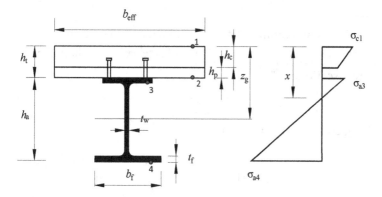

FIGURE 4.1 Schematic stress state of the HPCB subjected to elastic bending moment.

where
$A_a$ is the area of pre-engineered steel section
$h_t$ is the total height of HPCB
$h_a$ is the height of pre-engineered steel section
$h_c$ is the height of concrete slab
$h_s$ is the height of the centre of gravity of the steel section
$b_{eff}$ is the effective width of the concrete flange
$n$ is the ratio of composite material ($E_s/E_{cm}$)
$E_s$ is the elastic modulus of structural steel (210 GPa)
$E_{cm}$ is the secant modulus of concrete (EN 1992-1-1)

Depth of elastic neutral axis:

$$X_{ENA} = \frac{A_a\left(h_t + h_a - h_c - h_s\right) + \dfrac{b_{eff}h_c^2}{2n}}{A_a + \dfrac{b_{eff}h_c}{n}} \qquad (4.5)$$

Effective second moment of area:

$$I_{eff} = I_a + \frac{A_a\left(h_c + 2h_d + h_a\right)^2}{4\left(1 + nr\right)} + \frac{b_{eff}h_c^3}{12n} \qquad (4.6)$$

where

$I_a$ is the second moment of pre-engineered steel section

$r$ is the ratio of cross-sectional area of steel section relative to concrete section, $\dfrac{A_a}{b_{eff} h_c}$

Using parallel axis theorem:

$$I_a = \left[ \frac{b_{f1} t_{f1}^3}{12} + b_{f1} t_{f1} \bullet \left( h_s - \overline{x}_{f1} \right)^2 + \frac{t_w h_w^3}{12} + t_w h_w \bullet \left( h_s - \overline{x}_w \right)^2 \right. $$
$$\left. + \frac{b_{f2} t_{f2}^3}{12} + b_{f2} t_{f2} \bullet \left( h_s - \overline{x}_{f2} \right)^2 \right]$$

where

$b_{f1}$ is the top flange width of pre-engineered steel section

$t_{f1}$ is the top flange thickness of pre-engineered steel section

$h_w$ is the height of web of pre-engineered steel section

$t_w$ is the thickness of web of pre-engineered steel section

$b_{f2}$ is the bottom flange width of pre-engineered steel section

$t_{f2}$ is the bottom flange thickness of pre-engineered steel section

$h_s$ is the height of the centre of gravity of the pre-engineered section

$\overline{x}$ is the distance between $h_s$ and centre of gravity of element

## 4.3 CALCULATION OF DEFLECTIONS AND SERVICEABILITY STRESS

It is not specified in the EN 1994-1-1 deflection limits for composite beams; however, reference should be made to the three components of deflection defined in EN 1990. EN 1994-1-1 states that elastic analysis should be used to determine the deflection of composite members.

Deflection for four-point bending:

$$\delta = \frac{Pa \left[ 3L^2 - 4a^2 \right]}{48 E I_{\text{eff}}} \tag{4.7}$$

where

    $P$ is the point load

    $L$ is the total length of beam

    $a$ is the distance between the support and point load

    $E$ is the elastic modulus of the composite beam

    $I_{eff}$ is the effective second area of moment of the composite beam

Deflection for uniformed distributed loading:

$$\delta = \frac{5wL^4}{384EI_{eff}} \tag{4.8}$$

where

    $w$ is the uniformed distributed loading

    $L$ is the total length of beam

EC4 does not specify the deflection limits for composite beams. The deflection limits should be specified for individual projects depending on the sensitivity of the finishes, visual appearance, etc. to meet the client's needs. For composite slabs, the deflection due to live load should be limited to $L/250$. For propped construction, this deflection should also include the deflection due to prop removal.

Serviceability stress verifications are not prescribed in EC4. However, the guidance given in EN 1990, A1.4.2 may be applied to check the stress as part of the serviceability criteria where there is a risk of inelastic deflection under service loading. As deflections are based on elastic analysis, it seems prudent to have some verification that this assumption is valid.

## 4.4 VIBRATION CONSIDERATIONS

Vibration is mostly associated with the displacement of mass [3]. Hence, it is possible to categorize any vibration problem into two distinct groups: continuous systems and discrete systems. Continuous systems refer to systems where all the included mass elements are interconnected, as observed in scenarios like bending beams or guitar strings. Discrete systems refer to systems in which the masses involved are autonomous, as shown by the horizontal vibration of multi-story structures, where the floors represent the masses, and the columns represent the springs. Problems pertaining to continuous systems are often addressed by the application of integration techniques on continuous functions, whereas discrete systems can be effectively tackled by employing matrix operations. The second strategy, being comparatively easier, involves the discretization of continuous systems to enable their solution using methods designed for discrete systems. Among these techniques, finite element analysis is widely recognized as the most prominent.

The vibration consideration for HPCB adopts discrete systems, which are generally modelled from three components: point masses, springs and dampers. It can be determined by considering the forces applied on each mass by the other components and thus finding and solving the equations that link the velocity, acceleration and displacement to the external forces.

Frequency determination:

$$f_n = \frac{18}{\sqrt{\delta_w}} > 4\,\text{Hz} \tag{4.9}$$

where
Modal mass:

$$L_{\text{eff}} = 1.09(1.10)^{n_y-1}\left(\frac{EI_b}{mbf_0^2}\right)^{\frac{1}{4}} \tag{4.10}$$

$$S = \eta(1.15)^{n_x-1}\left(\frac{EI_b}{mbf_0^2}\right)^{\frac{1}{4}} \tag{4.11}$$

$$M = mL_{eff}S \tag{4.12}$$

where

$L_{eff}$ is the effective floor length

$EI_b$ is the dynamic flexural rigidity of secondary beam

$m$ is the distributed mass

$b$ is the floor beam spacing

$f$ is frequency

$S$ is the effective floor width

$n$ is the mode number

$M$ is modal mass

Floor response (Sum of peaks):

$$a_{w,rms} = \mu_e \mu_r \frac{0.1Q}{2\sqrt{2}M\xi} W \rho \tag{4.13}$$

where

$a_{w,rms}$ is the frequency-weighted rms acceleration

$\mu_e$ is the unity normalized amplitude at excitation point

$\mu_r$ is the unity normalized amplitude at response point

$Q$ is the static force exerted by "average person"

$\xi$ is the critical damping factor

$W$ is the weight factor

$\rho$ is the resonance build-up

Response factor:

$$R = \frac{a_{w,rms}}{0.005} \tag{4.14}$$

here

$a_{w,rms}$ is the frequency-weighted rms acceleration

TABLE 4.1   Vibration response limit

| Building Type | Min Vibration Requirements | Max $R$ |
|---|---|---|
| Semi-conductor industrial (high end) | VC-D | 0.625 |
| Semi-conductor industrial (low end) | VC-B | 0.25 |
| Office | ISO for office | 4 |
| Production | ISO for workshop | 8 |

The response limit is shown in Table 4.1 for the different types of industrial buildings. It shall be noted that this table is adopted by the JTC for industrial buildings in Singapore. The table adopted the vibration criterion (VC) based on International Standard of Organisation (ISO) guidelines for effects of vibration on people in buildings.

## 4.5  RECOMMENDATION ON FIRE PROOFING OF BEAMS

Most steel concrete composite beams are conventional downstand beams. The effects of fire on the steel sections are not much different from non-composite beams. Therefore, in most cases, steel sections need to be fire protected using intumescent coatings, sprayed non-reactive coating or boarding. In order to determine the right thickness of fire protection, the fire resistance period as well as the critical temperature and section factor required for the building must be known. Note that the critical temperature depends on the combination of actions which the beam is subjected to and the variation of the bending resistance with temperature. As far as fire design is concerned, the resistance of the beam under ultimate limit state is the only performance that needs to be evaluated. If the load to be supported by the beam in fire condition is unknown, the critical temperature may be determined as the temperature at which the resistance is equal to the design load. The required thickness of fire protection may then be evaluated based on how long it will take the steel section to heat up to this temperature [4].

## LIST OF SYMBOLS

| | |
|---|---|
| $A_a$ | is the area of pre-engineered steel section |
| $a$ | is the distance between the support and point load |
| $b$ | is the floor beam spacing |
| $b_{eff}$ | is the effective width of the concrete flange |
| $b_{f1}$ | is the top flange width of pre-engineered steel section |
| $b_{f2}$ | is the bottom flange width of pre-engineered steel section |
| $E$ | is the elastic modulus of the composite beam |
| $E_{cm}$ | is the secant modulus of concrete |
| $EI_b$ | is the dynamic flexural rigidity of secondary beam |
| $E_s$ | is the elastic modulus of structural steel |
| $f$ | is frequency |
| $h_a$ | is the height of pre-engineered steel section |
| $h_c$ | is the height of concrete slab |
| $h_d$ | is the height of the profile steel sheeting |
| $h_s$ | is the height of the centre of gravity of the steel section |
| $h_t$ | is the total height of HPCB |
| $h_w$ | is the height of web of pre-engineered steel section |
| $I_a$ | is the second moment of pre-engineered steel section |
| $I_{eff}$ | is the effective second area of moment of the composite beam |
| $L$ | is the total length of beam |
| $L_{eff}$ | is the effective floor length |
| $M$ | is modal mass |
| $m$ | is the distributed mass |
| $n$ | is the ratio of composite material/the mode number |
| $P$ | is the point load |
| $Q$ | is the static force exerted by "average person" |
| $S$ | is the effective floor width |
| $t_{f1}$ | is the top flange thickness of pre-engineered steel section |
| $t_{f2}$ | is the bottom flange thickness of pre-engineered steel section |
| $t_w$ | is the thickness of web of pre-engineered steel section |

| | |
|---|---|
| $W$ | is the weight factor |
| $w$ | is the uniformed distributed loading |
| $\bar{x}$ | is the distance between $h_s$ and centre of gravity of element |
| $a_{w,rms}$ | is the frequency-weighted rms acceleration |
| $\mu_e$ | is the unity normalized amplitude at excitation point |
| $\mu_r$ | is the unity normalized amplitude at response point |
| $\xi$ | is the critical damping factor |
| $\rho$ | is the resonance build-up |

## REFERENCES

1. Rackham, J. W.. AD 346: Design action during concrete for beams and decking in composite floor Advisory Desk, *New Steel Construction* 18(6), June 2010.
2. BSI. BS EN 1990 Eurocode 0: Basis of structural design, British Standards Institution (2002).
3. Smith, A. L., Hacks, S. J., Devine, P. J. SCI P354: Design of floors for vibration: a new approach The Steel Construction Institute (2009).
4. Simms, W. I., Hughes, A. F. SCI P359: Composite design of steel framed buildings The Steel Construction Institute (2011).

# CHAPTER 5

# Validation of Analysis Methods with Measurements and Finite Element Analysis

## 5.1 INTRODUCTION

To validate these derived design equations based on EC4 and analytical model established for high-performance pre-engineered steel concrete composite beam (HPCB) (details can be found in Chapters 3 and 4), two full-scale HPCB specimens are tested till failure in a four-point bending setup. The failure mode, bearing capacity, deflection, strain development and ductility of the beams are discussed in detail. Besides, a 3D nonlinear Finite Element (FE) model is developed and validated against the test results. This FE model provides more insight on the development

DOI: 10.1201/9781032626932-5

and distribution of stress and deformation during loading. Aside from creating track record that facilitates industry adoption, deep understanding of the actual behaviour of the HPCB would allow the practitioners to further optimize the design and advance productivity and economic benefits.

## 5.2 TEST SETUP AND PROCEDURES

### 5.2.1 Specimen Design

Two full-scale HPCBs replicating the primary and secondary beams of the next-generation industry buildings were fabricated, as shown in Figure 5.1 (a and b), respectively. The detailed design process can be found in Appendix A. The primary beam was fabricated to 7.4 m with a slab width of 1.5 m, to simulate the sagging portion of the 12 m long continuous beam, while the secondary beam was fabricated to full length of 12 m with a slab width of 1.4 m to simulate the effective simply supported secondary beam. Shear studs that were 19 mm diameter headed were welded to the top flange of the steel sections to ensure full shear connections between the concrete slab and the pre-engineered steel sections.

### 5.2.2 Material Properties

Mix design S1 (details are provided in Chapter 2) was adopted for the high-performance green concrete (HPGC) in grade C55/67. Compressive strength of 76.5 MPa was achieved in 28 days for the

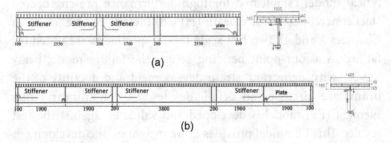

FIGURE 5.1  Tested full scale HPCB specimens (unit: mm). (a) Primary beam; and (b) Secondary beam.

same batch of concrete used for casting the slabs. The asymmetrical pre-engineered steel sections were fabricated from S460M structural steel plates (manufactured to EN 10025-4) by welding. Plates in three thickness, that is, 8-, 12- and 14-mm thick plates, were used for top flange, web and the bottom flange, respectively. The profile steel sheeting with a thickness of 1 mm was placed underneath the concrete slab as a permanent formwork. Header shear studs with a Ø19 shank were used to provide the composite action. The material properties of steel plate and steel bars are summarized in Table 5.1.

### 5.2.3 Test Setup and Loading Scheme

Figure 5.2 presents the experimental setup of the static load test, using the secondary beam as an example. The HPCBs are tested under a four-point bending configuration with a pure bending region of 1,900 and 3,800 mm for the primary and secondary beams, respectively. Compared to three-point bending, the four-point bending configuration produces peak stresses along an extended region at the midpoint. This configuration also eliminates the shear stress between the two points of loading, creating a pure bending section which is favourable for flexural behavioural study. To prevent out-of-plane buckling/torsion during the test, eight lateral braces are symmetrically installed near the supports and loading points, as shown in Figure 5.2.

TABLE 5.1   Mechanical properties of materials

| Category | Thickness/ Diameter (mm) | Yield Strength (MPa) | Elastic Modulus (GPa) |
|---|---|---|---|
| Steel Beam | 8 (top flange) | 595.7 | 216.08 |
| | 12 (web) | 585.3 | 215.82 |
| | 14 (bottom flange) | 482.1 | 211.18 |
| Rebar | 10 (longitudinal) | 439.2 | — |
| | 12 (transverse) | 539.8 | — |

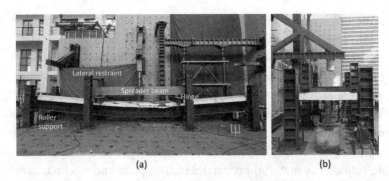

FIGURE 5.2   Test setup for the secondary beam. (a) Elevation view; and (b) Side view.

Step-loading approach was adopted for the experimental study. To calibrate the test setup, a pre-loading of 50 kN with a loading rate of 1 kN/s is applied to the specimen at each step until the loading approaches the designed linear elastic limit. During the official loading, the load is applied to the HPCB specimen via force-control with a loading rate of 1 kN/s during the linear elastic stage. Upon entering the plastic stage, the loading process shifted from force-controlled to displacement-controlled with a loading rate of one mm/min. At every interval of five minutes, the load is maintained and held constant for a duration of two minutes when cracks formation and development on the HPGC slab are recorded. By monitoring the crack propagation under the sustained load, the structural integrity and behaviour of HPCB at different load levels can be assessed. Loading is stopped when the specimen fails with major concrete crushing, and the load is subsequently reduced to 0.85 of the peak load for recoding purpose before unloading.

To monitor the vertical displacement during loading, three linear variable displacement transducers (LVDT) are installed at the bottom of the pre-engineered section. Four more LVDTs are installed at both ends of the specimen to capture any longitudinal movement during the test. Strain gauges are attached along the cross-sectional height of the pre-engineered section to capture the strain history during the test, with the aim to examine the movement of neutral axis from elastic to plastic behaviour. Besides the pre-engineered section, several strain gauges are also attached to the top surface and along the cross-sectional height of the HPGC slab to acquire and record the strain history during the test. The details of instrumentation are shown in Figure 5.3. In addition to the strain gauges, a high-resolution digital camera is employed to capture the strain, displacement and crack propagation histories of the pre-engineered section by using the Digital Image Correlation (DIC) technique.

### 5.2.4 Test Results and Discussion

#### 5.2.4.1 Failure Modes

It was observed that during the linear elastic stage, deflection of the primary HPCB increased gradually with the inclination angle at both ends of the composite beam. The strain distribution along the web of the steel section remained linear as indicated by the data logging system. As the load reached 1,108 kN (equivalent to 1,468 kNm, 82% of the linear elastic limit), two longitudinal cracks were observed at the mid-span of the HPGC slab and slowly propagated in the transverse direction towards the edges. While the sound of concrete cracking was noticed, load-bearing capacity was not apparently affected. As the loading increased to 1,300 kN (1,722 kNm), the bottom flange of the pre-engineered section began to yield (0.24% strain), indicating that the HPCB had entered the elastoplastic stage (elastic-plastic stage). It is observed that the strain on the web section close to the bottom flange has yielded similarly, while a large portion of the web is still undergoing relatively small strains of 0.05–0.17%, as compared to the bottom flange. It is also observed that the profile steel sheeting has been separated from the concrete slab with a gap width of approximately 3 mm at the mid-span. As the load continued to increase and reached 1,430 kN (1,895 kNm,

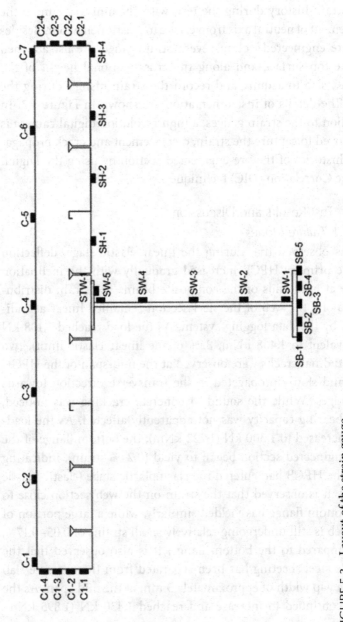

FIGURE 5.3  Layout of the strain gauge.

106% of the linear elastic limit), the recorded vertical displacement reached 27.4 mm. It was observed that approximately a third of the web had yielded (strain ranged from 0.21 to 0.26%), and multiple small cracks appeared on the side of the HPGC slab.

The separation between the HPGC slab and profile steel sheeting became increasingly obvious as the load increased further, and multiple cracks with lengths ranging from 200 mm to 600 mm emerged on the HPGC slab. Several through-thickness cracks occurred when the load increased to 2,040 kN (2,703 kNm, 122% of the design plastic load). A circular bulge with diameter of approximately 50 mm appeared on the top surface of the HPGC slab as the load reached 2,154 kN (2,854 kNm, 129% of the design plastic load). At the peak load of 2,328 kN (3,083 kNm, 140% of the design plastic load), the recorded mid-span deflection reached 113.3 mm, and extensive cracking was observed on the top surface of the concrete slab. Finally, an extensive area of the HPGC slab was crushed at the pure bending region. At this stage, the concrete slab had essentially lost its load-carrying capacity and the steel section buckled as a result. The final deformation of the primary HPCB is presented in Figure 5.4. Noted that after the test, the concrete slab was demounted. No obvious deformation in the studs was found, as shown in Figure 5.5. As a result of concrete crushing, significant buckling was observed at the top flange of the steel section.

Similar to the primary HPCB, during the linear elastic stage, both HPCB slab and pre-engineered steel section of the secondary HPCB remained intact. However, six small cracks were observed

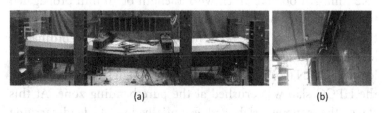

(a)                                                        (b)

FIGURE 5.4  Failure mode of the primary HPCB. (a) Concrete slab crushing; and (b) Buckling of steel section.

FIGURE 5.5    Deformation of the top flange after test.

along the longitudinal axis on the top surface of the HPGC slab as the load reached 390 kN (780 kNm, 53% of the linear elastic limit). Two cracks were formed on the right side of the HPCB with a length of 1,900 and 2,500 mm. Two more were formed on the left side with a length of 1,800 and 2,300 mm. The last two cracks were found in the middle portion with lengths of 500 and 1,300 mm. Formation of new cracks and propagation of existing cracks were observed, as the load increased. At 550 kN (1,100 kNm, 75% of the linear elastic limit), obvious deformation was observed when the vertical displacement recorded in the mid-span reached 47.2 mm. As the load increased to 703 kN (1,406 kNm, linear elastic limit), it is observed that the bottom flange (0.24% strain) together with the bottom web (0.22% strain) of the pre-engineered section were yielded.

As the loading proceeded to increase to approximately 800 kN (1,600 kNm, 76% of the design plastic resistance), a third of the web yielded (0.27–0.22% strain) and the displacement reached 75.2 mm. An oblique crack with a length of 70 mm propagated from one loading beam to the other. The profile steel sheeting and the HPGC at this stage were not able to deform co-ordinately, leading to the separation between the two elements. At the peak load, 1,328 kN (2,656 kNm, 126% of the design plastic resistance), the HPGC slab was crushed at the pure bending zone. At this stage, the concrete slab had essentially lost its load-carrying capacity, which resulted in a sudden drop in moment-displacement curve. The recorded vertical displacement was 353.1 mm at this moment. Figure 5.6 presents the final deformation of secondary HPCB.

### 5.2.4.2 Bearing Capacity

The moment-displacement curve, together with the theoretical linear elastic displacement and designed plastic moment resistance, for mid-span of the primary and secondary HPCB are plotted in Figures 5.7 and 5.8, respectively. In Figure 5.7, it is evident that when the primary HPCB was loaded to the serviceability limit state (SLS) of 930 kNm, with a corresponding deflection at mid-span of 20 mm, the behaviour is still within the linear elastic stage. The specimen yielded at approximately 1,653 kN, after which the specimen has undergone irreversible deformation and the formation of cracks. At the peak moment of 3,083 kNm, which is 33.5% higher than the designed plastic resistance, the HPGC slab was crushed at the pure bending zone with a deflection of

FIGURE 5.6    Failure mode of the secondary HPCB.

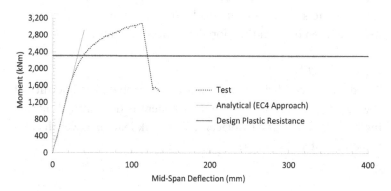

FIGURE 5.7    Bending moment-deflection curves of the primary HPCB.

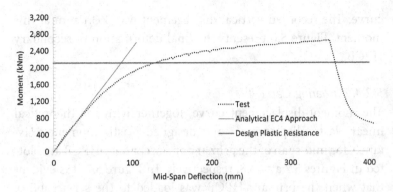

FIGURE 5.8   Bending moment-deflection curves of the secondary HPCB.

113 mm at mid-span. As a result of the crushing, the concrete slab has essentially lost its load-carrying capacity with an indication of sudden drop in the moment-deflection curve.

Similarly for the secondary HPCB specimen (moment-displacement curve is shown in Figure 5.8), before the specimen is loaded to approximately 1,400 kNm, the specimen is still in the linear elastic stage. Afterwards, the moment-displacement curve shifted to the nonlinear/yield stage, and the deflection became more pronounced as the moment continued to increase. It reached the peak moment of 2649.4 kNm, which is 25.3% higher than the designed plastic moment resistance when the HPGC slab is crushed with a deflection recorded at 350.2 mm. At this moment, the concrete slab has lost its load-carrying capacity and resulted in the sudden drop in the moment-deflection curve.

### 5.2.4.3 Ductility

The ductility coefficient index is used to evaluate the ductility of the beam specimens [1–4]. This coefficient is determined by taking the division of the displacement at peak load ($\delta_u$) over the displacement at yield ($\delta_y$) [4].

$$u = \frac{\delta_u}{\delta_y} \tag{5.1}$$

TABLE 5.2    Displacement ductility coefficient of the specimens

| Specimen | Mid-Span Displacement at Yielding | Mid-Span Displacement at Failure | Ductility Coefficient |
|---|---|---|---|
| Primary HPCB | 23.9 mm | 113.3 mm | 4.74 |
| Secondary HPCB | 63.1 mm | 353.1 mm | 5.60 |

The ductility coefficients of these two HPCB specimens are presented in Table 5.2. In general, the displacement ductility coefficient of a ductile beam should be no less than 3.0 [5]. As can be seen in Table 5.2, in this study, both specimens showed ductility coefficients of more than 4.5, suggesting good ductility. The results also imply that the ductility coefficient increases with the effective span of the beam.

### 5.2.4.4 Strain Development in the Web

Figure 5.9 illustrates the strain development in the web of the primary HPCB captured by DIC, where $P_u$ is the peak load. As present in Figure 5.9, it is observed that strain propagated upward from the bottom and generally linear proportional through the height. At 0.2 $P_u$, the web section is observed to be in the elastic stage. As the load increased from 0.4 to 0.6 $P_u$, it is evident that the bottom web has started to yield. When subjected to 0.8 $P_u$, a substantial portion of the bottom web has yielded.

Figure 5.10 compares the load-strain curve obtained from DIC and test strain gauge measurement for the middle section of the web (SW-3, indicated in Figure 5.3). In Figure 5.10, it is evident

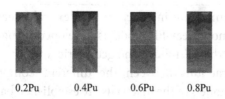

0.2Pu        0.4Pu        0.6Pu        0.8Pu

FIGURE 5.9    Web strain development in the middle span of the primary HPCB.

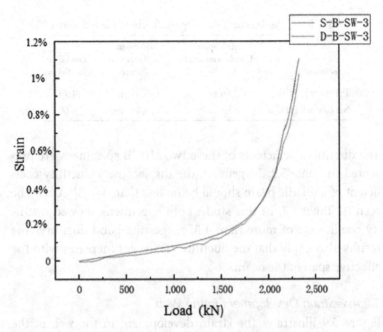

FIGURE 5.10 Load-strain curve obtained from DIC compared to test strain gauge measurements.

that the two curves agree well with each other, suggesting high accuracy of strain data obtained from DIC. It is also observed that the mid portion of the web is in the linear elastic stage till 1,300 kN. As the load continued to increase, high accuracy of plastic strain measurement was observed till the specimen was loaded until the peak load.

## 5.3 FINITE ELEMENT ANALYSIS

In order to gain more insights on the flexural behaviour of the HPCB, a 3D nonlinear FE model is developed using FE package ABAQUS. In the FE model, the geometric and material nonlinearities, interactions between the different components, and constitutive models of the materials are established based on compression and tensile test results. The actual loading and boundary conditions used in the tests are reproduced with necessary simplifications to minimize the file size of the model. The developed model is finally validated against the actual test results.

## 5.3.1 General Information of the FE Model

The FE model consists of five primary elements, that is, the pre-engineered HPS section, HPGC slab, profile steel sheeting, headed shear studs and reinforcement steel bars in the longitudinal and transverse directions, as presented in Figure 5.11. All these elements, except for the HPGC slab and reinforcement bars, are modelled using four-node reduced shell elements (S4R). The HPGC slab is modelled using an eight-node hexahedral solid element (C3D8R), while the reinforcement bars are modelled with a two-node linear truss model (T3D2). A sweep meshing option with typical meshes is adopted, as shown in Figure 5.12. A mesh-sensitive analysis is also carried out. It turns out that 50-mm element dimension is appropriate for this analysis. The contact interactions used for the five parts are listed in Table 5.3

FIGURE 5.11   Elevation section of the FE model.

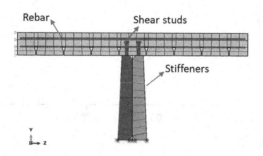

FIGURE 5.12   Cross section of the FE model.

TABLE 5.3 Interaction between different elements

| Item 1 | Item 2 | Interaction Behaviour |
| --- | --- | --- |
| Concrete slab | Steel sheet | Tie |
| Concrete slab | Studs | Embedded |
| Concrete slab | Bars | Embedded |
| Concrete slab | Load cells | Contact, with friction coefficient of 0.3 |
| Steel beam | Steel sheet | Contact, with friction coefficient of 0.3 |
| Steel beam | Studs | Tie |

## 5.3.2 Material Model

### 5.3.2.1 Concrete Model

The concrete-damaged plasticity model is adopted for the HPGC. The compressive stress–strain relationship is depicted in Figure 5.13, which is generated using the cube strength and the equation provided in the FIB Model Code [6]:

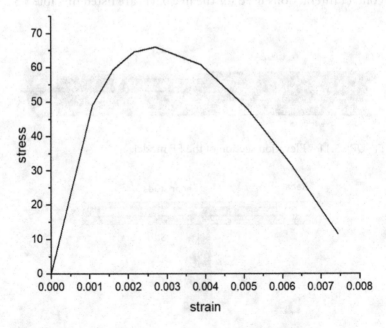

FIGURE 5.13 Compressive stress-strain curve for concrete.

$$\frac{\sigma_c}{f_{cm}} = -\left\{\frac{k\cdot\eta-\eta^2}{1+(k-2)\cdot\eta}\right\} \text{for } |\varepsilon_c| < |\varepsilon_{c,\lim}| \qquad (5.2)$$

The tensile strength $f_{ctm}$ of the concrete and fracture energy $G_f$ can be stimulated using the equations suggested by FIB Model Code [6]

$$f_{ctm} = 2.12\cdot\ln(1+0.1f_{cm}) \qquad (5.3)$$

$$G_f = 73\cdot f_{cm}^{0.18} \qquad (5.4)$$

### 5.3.2.2 Steel Model

The stress–strain model used for the reinforcement bar, pre-engineered HPS section, and profile steel sheeting is presented in Figure 5.14. Note that the ratio of hardening modulus to elastic modulus ratio for reinforcement bars (0.03) and structural steel (0.02) differs slightly in the hardening stage [7]. As no obvious plastic deformation was observed in any of the spreader beams or plates during the test, linear elastic material model is used for those elements.

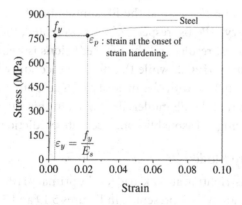

FIGURE 5.14   Stress-strain relationship for steel.

## 5.4 COMPARISON WITH MEASUREMENTS OBTAINED FROM EXPERIMENTAL STUDY

### 5.4.1 Bearing Capacity

The moment-displacement curves predicted by the FE modelling for both primary and secondary beams are shown in Figures 5.15 and 5.16, respectively. From Figure 5.15, it can be seen that the FE model produced highly accurate moment-displacement curve for the primary HPCB. The FE model accurately predicted not only the linear elastic stage but also the peak load. In addition, a detailed analysis of the FE stress distribution also shows the yielding of pre-engineered HPS section when the concrete crushed.

Similarly in Figure 5.16, the FE model agreed well with the analytical model in the linear elastic stage and test result. However, the simulation results are approximately 8% higher than the test curve at the hardening stage. This could be a result from the difficulty in reproducing the construction junctions of HPGC from the actual secondary HPCB to the FE model. Nonetheless, the FE model accurately predicted the peak load and the failure mode of concrete crushing afterwards. Under a detailed analysis of the FE stress distribution, significant yielding of the pre-engineered high-performance steel section is noticed when the concrete slab is crushed.

Figures 5.17 and 5.18 compare the moment resistance of the experimental results against the FE simulations for the primary and secondary HPCBs, respectively. As presented, the difference between the test results and the FE predictions is well within 5% for the primary HPCB, while that for the secondary beam is in general within 10% with the majority of data points within 5%. This implies that the FE models developed in this study are capable of delivering a reasonably good accurate prediction.

### 5.4.2 Failure Mode

The stress distribution at peak load for the primary HPCB obtained from the FE analysis is presented in Figures 5.19 and 5.20. The FE

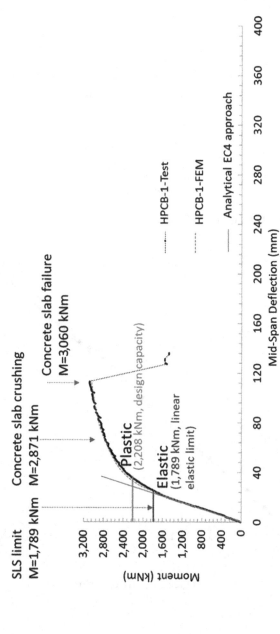

FIGURE 5.15   Validation of FE model for the primary HPCB.

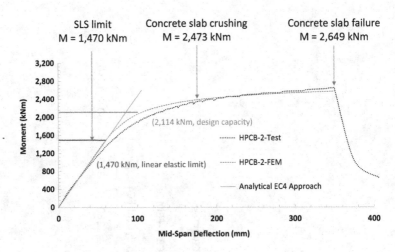

FIGURE 5.16   Validation of FE model for the secondary HPCB.

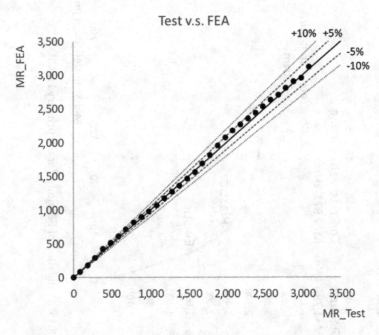

FIGURE 5.17   Moment resistance comparison-primary HPCB.

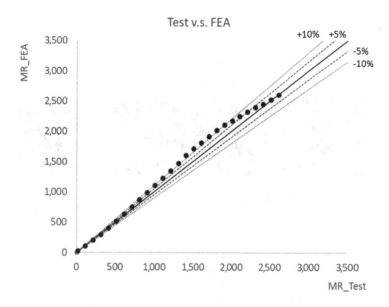

FIGURE 5.18   Moment resistance comparison-secondary HPCB.

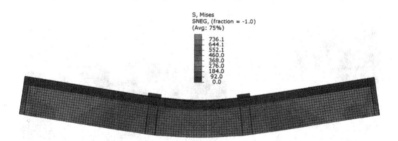

FIGURE 5.19   Stress distribution in the primary HPCB at the failure stage.

models exhibited a similar flexural behaviour and failure mode to the actual experimental HPCB specimens (Figure 5.4). Cracks on the top HPGC were mainly longitudinal along the axis of the composite beam, as shown in Figure 5.21. When the specimen reached its design bending capacity, a large area of pre-engineered high-performance steel section was yielded.

Figure 5.22 compares the web strain distribution at the failure stage of primary HPCB. Figure 5.22 a shows the strain distribution

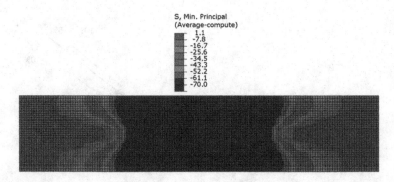

FIGURE 5.20   Stress distribution in the concrete slab

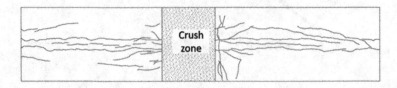

FIGURE 5.21   Crack propagation in the concrete slab.

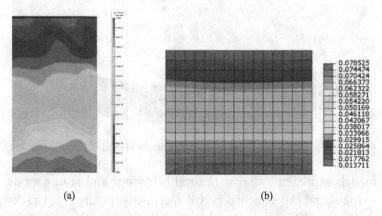

(a)                                         (b)

FIGURE 5.22   Web strain distribution comparison. (a) Web strain obtained from DIC; and (b) Web strain obtained from FEM.

obtained from DIC, and Figure 5.22b shows the strain obtained from FE analysis. As can be seen from these two figures, the results from DIC and FE analysis agree well with each other. At the failure stage, the entire web was yielded. FE analysis suggests that the maximum tensile strain of the web near the bottom flange is around 7.85%.

It can be concluded that the FE analysis developed in this study could be employed as a reliable tool for investigating the flexural resistance of HPCB.

## 5.5 FURTHER STUDY ON THE MOMENT–NEUTRAL AXIS RELATIONSHIP

In this section, the relationship between moment versus location of neutral will be discussed in further detail. It is known that the location of the neutral axis shifts from elastic to plastic as the load increases, as shown in Figure 5.23. However, in the rigid plastic model, only the rectangular stress block is considered for determining the plastic neutral axis (PNA) at the full plastic stage.

From the recorded strain history, the strain distribution along the beam depth is plotted in Figure 5.24, at different levels of loading from $0.2P$ to $P$ ($P$ is the peak load). The locations of the neutral axis of the beams were determined through linear extrapolation from the recorded strain of the pre-engineered structural steel sections.

Through Figure 5.24, the locations of the neutral axis against the moment can be determined and plotted out in Figure 5.25. When examining the locations of the neutral axis, it is observed that the moment–neutral axis relationship is nicely fitted into a regression model through a polynomial equation of an order of 5, with a coefficient of determination value of 0.99. It can be seen from Figure 5.25 that the moment–neutral axis relationship gradually shifts upward within the steel section, which greatly increases the moment capacity. However, the neutral axis has a drastic upward progression when it is in the profile steel sheeting, accompanied by a slight increase in the moment capacity. It is rationale to deduct that the great increase in the moment during the early stage is due to the yield of the lower pre-engineered section. At this stage, the steel section has reached approximately

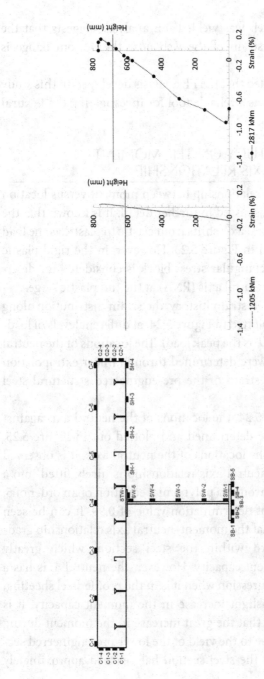

FIGURE 5.23  Strain distribution under elastic and failure stages.

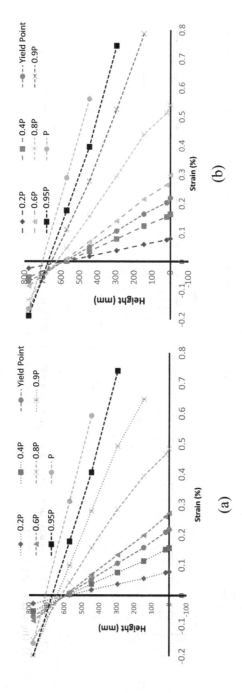

FIGURE 5.24 Strain distribution during loading.

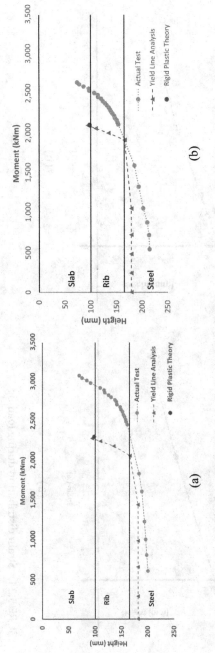

FIGURE 5.25 Neutral axis position – moment relationship (yield line analysis v.s. tes results). (a) The primary HPCB; and (b) The secondary HPCB.

90% of the plastic resistance. As the lower section is further away from the neutral axis, the effects of the long lever arm are maximized.

Note that the actual PNA is 4 mm higher than the theoretical value determined by the rigid plastic model based on the EC4 approach. It is likely due to the fact that the tensile stress in the flange and strain hardening behaviour is not taken into consideration, which is expected to have beneficial effects on the moment capacity.

## 5.6 SUMMARY

Guidance for testing HPCBs including setup, test procedure and data collection and interpretation is given. The use of digital image correlation techniques is also discussed in comparison with conventional strain gauging methods. The failure mode, bearing capacity, deflection, strain development and ductility of the beams are discussed in detail. Besides, a 3D nonlinear Finite Element (FE) and yield line model are developed and validated against the test results. This FE model provides more insight on the development and distribution of stress and deformation during loading. Aside from creating track record that facilitates industry adoption, deep understanding of the actual behaviour of the HPCB would allow the practitioners to further optimize the design and advance productivity and economic benefits.

The main conclusions are as follows:

(1) Both high-performance steel-concrete composite beams show typical ductile flexural failure modes with bottom steel flange yielded and concrete slab crushed. No obvious bending deformation in the studs is found, indicating that the studs provided sufficient shear resistance. Comparing the bearing capacity obtained from the test with the EC4 approach, it is evident that the design plastic resistance recommended by EC4 is still conservative. The actual plastic moment resistance is 33.5% higher than the design plastic

moment resistance for the primary HPCB and 25.3% higher for the secondary HPCB.

(2) FE analysis introduced in this study is able to produce a reasonably accurate prediction (5–10% deviation from the test results). In addition, the analytical model is highly accurate in predicting the deflection in the linear elastic stage, with a difference of 0.18 and 0.96 mm for the primary and secondary HPCBs at the serviceability limit state, respectively.

(3) Monitoring of strain development in the web of the steel section reveals that the neutral axis gradually shifts up from steel section into concrete slab. To predict the shift of neutral axis, which is valuable for understanding the elastic and plastic behaviour of the HPCBs, the yield model developed in Chapter 3 is applied. It is shown that this model is able to predict the shift of the neutral axis accurately but is conversative for predicting the bending moment resistance.

## REFERENCES

1. Han, L. H. & Tao, Z. Ductility coefficient of concrete filled steel tubular columns with square sections. *Earthquake Engineering and Engineering Vibration* 20(4), 56–65 (2000).
2. Shen, P. S. *Difficult interpretation of tall building structures.* China Building Industry Press (2003).
3. Hassani, M., Suhatril, M., Shariati, M. & Ghanbari, F. Ductility and strength assessment of HSC beams with varying tensile reinforcement ratios. *Structural Engineering and Mechanics*, 48(6), 833–848 (2013).
4. Qiao, P. L., Xie, L. L. & Lv D. G. *Structural dynamics: Theory and its application in earthquake engineering.* Beijing: Higher Education Press (2007).
5. Xu, H. B. & Deng, Z. C. Experimental research on flexural behavior of prestressed ultra-high performance steel fiber concrete beams. *Journal of Building Structures* 35(12), 58–64 (2014).
6. FIB 2010, *Fib model code for concrete structures* (2010).
7. Tao, Z., Wang, X. Q. & Uy, B. Stress-strain curves of structural and reinforcing steels after exposure to elevated temperatures. *Journal of Materials in Civil Engineering* 25(9), 1306–1316 (2013).

# Worked Example

## *HPCB System for a 12 m × 12 m Grid Floor with Live Load of 20 kPa*

## A.1 STRUCTURAL ANALYSIS

The typical structural layout is presented in Figure A.1. The 12 m primary beam is connected to the column via fixed-end connection, while secondary beams are simply supported by the primary

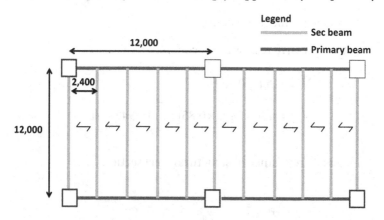

FIGURE A.1   Typical structural layout (12 m × 12 m grid).

beams, with 180 mm slab thickness. The profile steel sheeting used has a rib height of 54 mm with a centre-to-centre distance of 200 mm.

A structural analysis software was deployed to obtain the shear force and bending moment diagrams, as shown in Figure A.2. Design loads include the imposed live loads, self-weight, and imposed dead loads.

## A.2 ULTIMATE LIMIT STATE

### A.2.1 Design of the Secondary Beam

The preliminary dimensions shown in Figures A.3 and A.4 are determined using spreadsheet.

1. Effective section

$$
\begin{aligned}
b_{\text{eff}} &= b_0 \sum b_{\text{ei}} \\
b_{\text{ei}} &= \frac{L_e}{8} = \frac{12000}{8} = 1500 \\
b_{\text{eff}} &= 80 + 1500 + 1500 = 3000 > \text{bay width} \\
\therefore b_{\text{eff}} &= 2400 \text{ mm}
\end{aligned}
\tag{A.1}
$$

2. Plastic moment resistance

Compressive resistance of concrete

$$
\begin{aligned}
N_{c,f} &= h_c b_{\text{eff}} 0.85 f_{cd} \\
N_{c,f} &= 126 \times 2400 \times 0.85 \left( \frac{55}{1.5} \right) = 9424 \text{ kN}
\end{aligned}
\tag{A.2}
$$

Tensile resistance of structural steel section

$$
\begin{aligned}
N_{\text{pl.}a} &= \frac{A_a f_y}{\gamma_a} \\
N_{\text{pl.}a} &= \frac{8860 \times 460}{1} = 4075 \text{ kN}
\end{aligned}
\tag{A.3}
$$

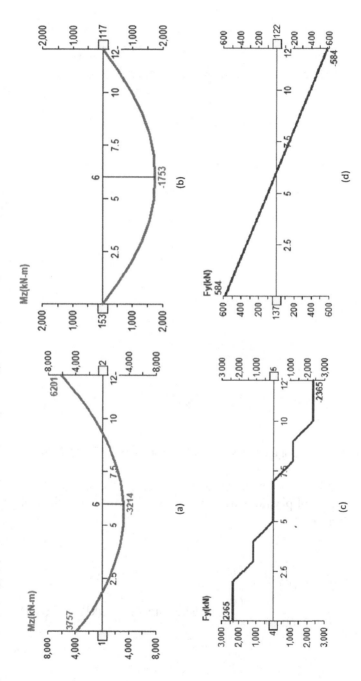

FIGURE A.2  Being moment and shear force diagram. (a) Bending moment of the primary HPCB; (b) Bending moment of the secondary HPCB; (c) Shear force of the primary HPCB; (d) Shear force of the secondary HPCB.

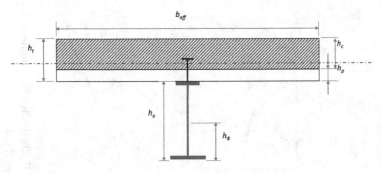

FIGURE A.3  Typical cross-section of secondary HPCB.

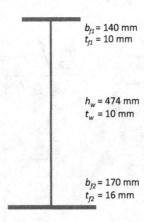

$b_{f1}$ = 140 mm
$t_{f1}$ = 10 mm

$h_w$ = 474 mm
$t_w$ = 10 mm

$b_{f2}$ = 170 mm
$t_{f2}$ = 16 mm

FIGURE A.4  Pre-engineered section details of the secondary HPCB.

Position of plastic neutral axis (PNA) from the top fibre of concrete .

$$x = \frac{N_{\mathrm{pl},a}}{b_{\mathrm{eff}}\,0.85\,f_{\mathrm{cd}}} < h_c$$

$$x = \frac{4075 \times 1000}{2400 \times 0.85 \times 36.7} = 54.5 \text{ mm} < h_c \therefore \text{Ok!} \qquad \text{(A.4)}$$

Pre-engineered section's centre of gravity

$$h_s = \frac{\sum(A_i y_i)}{A_a} = \frac{b_{f1}t_{f1}(h_a - 0.5t_{f1}) + h_w t_w(0.5h_w + t_{f2}) + 0.5b_{f2}t_{f2}^2}{A_a}$$

$$h_s = \frac{180 \times 10 \times \left(500 - \frac{10}{2}\right) + 474 \times 10 \times \left(\frac{10}{2} + 16\right) + \frac{170 \times 16^2}{2}}{8860}$$

$$= 216 \text{ mm} \tag{A.5}$$

Plastic moment resistance

$$\begin{aligned} M_{\text{pl,Rd}} &= N_{\text{pl},a}[h_a - h_s + h_c + h_p - 0.5x]M_{\text{pl,Rd}} \\ &= \left[ 9424 \times 1000 \left[ 500 - 216 + 126 + 54 - \frac{53.3}{2} \right] \right] \times 10^{-6} \\ &= 1779.9 \text{ kN} > M_{\text{Ed}} \therefore \text{Ok!} \end{aligned} \tag{A.6}$$

3. Design shear resistance

$$V_{\text{pl},a,\text{Rd}} = A_v \frac{f_{\text{yd}}}{\sqrt{3}}$$

$$V_{\text{pl},a,\text{Rd}} = \left[ 474 \times 10 \times \frac{460}{\sqrt{3}} \right] \times 10^{-3} = 1258.9 \text{ kN} > V_{\text{Ed}} \therefore \text{Ok!} \tag{A.7}$$

Shear buckling check

$$\frac{h_w}{t_w} \leq 72\varepsilon$$

$$\frac{474}{10} = 47.4 \leq \left( 72 \times \sqrt{\frac{235}{460}} \right) = 51.5 \therefore \text{Ok!} \tag{A.8}$$

4. Shear connection

A typical cross-section of shear stud and profile steel sheeting shown in Figure A.5 is used for designing the shear capacity.

Stud failure

$$P_{Rd} = \frac{0.8 f_u \pi d^2}{4\gamma_v} = 0.16 f_u \pi d^2$$
$$P_{Rd} = \left(0.16 \times 450 \times \pi 19^2\right) \times 10^{-3} = 81.7 \text{ kN} \qquad (A.9)$$

Concrete failure

$$P_{Rd} = \frac{0.29\alpha d^2 \sqrt{f_{ck}E_{cm}}}{\gamma_v} = 0.232\alpha d^2 \sqrt{f_{ck}E_{cm}}$$
$$\left(\frac{h_{sc}}{d}+1\right) = \left(\frac{95}{19}+1\right) = 5 > 4 \therefore \alpha = 1$$
$$P_{Rd} = \left[0.232 \times 1 \times 19^2 \sqrt{55 \times 38000}\right] \times 10^{-3} = 121 \text{ kN}$$
$$\therefore P_{Rd} = 81.7 \text{ kN} \qquad (A.10)$$

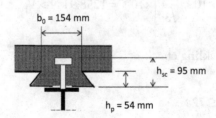

FIGURE A.5  Cross-sectional details of the shear stud and profile steel sheeting.

Reduction factor for transverse profile steel sheeting and welded through (2 studs per trough)

$$k_t = \frac{0.7}{\sqrt{n_r}} \frac{b_0}{h_p} \left( \frac{h_{sc}}{h_p} - 1 \right) \le k_{t,max}$$

$$k_{t,max} = 0.8$$

$$k_t = \frac{0.7}{\sqrt{2}} \frac{150}{54} \left( \frac{95}{54} - 1 \right) = 1.04 \le k_{t,max}$$

$$\therefore k_t = 0.8 \tag{A.11}$$

HPCB is based on full shear connection

$$\eta = \frac{N_{Rd}}{\min \left( N_{c,f}, N_{pl,a} \right)}$$

$$1 = \frac{N_{Rd}}{4075}$$

$$N_{Rd} = 4075 \ kN$$

Rows of shear studs required

$$no = \frac{N_{Rd}}{R_{Rd}}$$

$$no = \frac{4075}{81.7 \times 0.8 \times 2} \approx 31 nos$$

Spacing between shear studs

$$s = \frac{L_{eff}/2}{n-1}$$

$$s = \frac{12000/2}{31-1} = 200 \ mm$$

$$\therefore s = 200 \ mm$$

## A.2.2 Design of the Primary Beam

Overview of the HPCB flooring system is shown in Figure A.8 for example. The preliminary dimensions shown in Figures A.6 and A.7 are determined using spreadsheet.

1. Effective section

$$b_{eff} = b_0 \sum b_{ei}$$
$$b_{ei} = \frac{L_e}{8} = \frac{7200}{8} = 900$$
$$b_{eff} = 80 + 900 + 900 = 1880 < \text{bay width}$$
$$\therefore b_{eff} = 1800 \text{ mm} \tag{A.12}$$

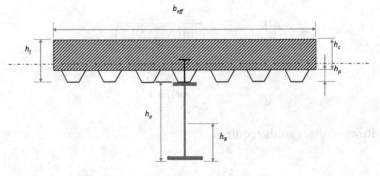

FIGURE A.6  Typical cross-section of primary HPCB.

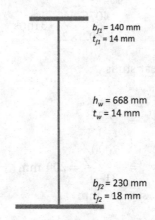

$b_{f1}$ = 140 mm
$t_{f1}$ = 14 mm

$h_w$ = 668 mm
$t_w$ = 14 mm

$b_{f2}$ = 230 mm
$t_{f2}$ = 18 mm

FIGURE A.7  Pre-engineered section details of the primary HPCB.

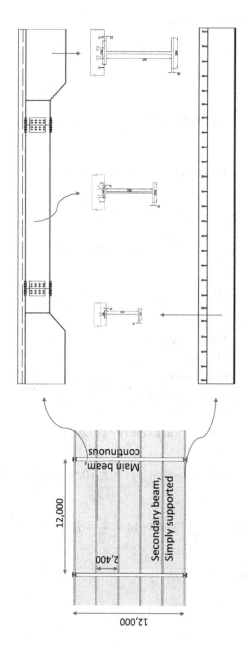

FIGURE A.8   Overview of the HPCB flooring system.

2. Plastic moment resistance

Compressive resistance of concrete

$$N_{c,f} = h_c b_{\text{eff}} 0.85 f_{cd}$$

$$N_{c,f} = 126 \times 1880 \times 0.85 \left( \frac{55}{1.5} \right) = 7382.8 \text{ kN} \qquad \text{(A.13)}$$

Tensile resistance of structural steel section

$$N_{\text{pl}.a} = \frac{A_a f_y}{\gamma_a}$$

$$N_{\text{pl}.a} = \frac{15452 \times 460}{1} = 6798.9 \text{ kN} \qquad \text{(A.14)}$$

Position of PNA

$$x = \frac{N_{\text{pl},a}}{b_{\text{eff}} 0.85 f_{cd}} < h_c$$

$$x = \frac{6798.9 \times 1000}{1880 \times 0.85 \times 36.7} = 116 \text{ mm} < h_c \therefore \text{Ok!} \qquad \text{(A.15)}$$

Pre-engineered section's centre of gravity

$$h_s = \frac{\sum (A_i y_i)}{A_a} = \frac{b_{f1} t_{f1} \left( h_a - 0.5 t_{f1} \right) + h_w t_w \left( 0.5 h_w + t_{f2} \right) + 0.5 b_{f2} t_{f2}^2}{A_a}$$

$$h_s = \frac{140 \times 14 \times \left( 700 - \dfrac{14}{2} \right) + 668 \times 14 \times \left( \dfrac{14}{2} + 18 \right) + \dfrac{230 \times 18^2}{2}}{15452}$$

$$= 303.4 \text{ mm} \qquad \text{(A.16)}$$

Plastic moment resistance

$$M_{pl,Rd} = N_{pl,a}[h_a - h_s + h_c + h_p - 0.5x]M_{pl,Rd}$$

$$= \left[ 6798.9 \times 1000 \left[ 700 - 303.4 + \right. \right.$$

$$\left. \left. 126 + 54 - \frac{116}{2} \right] \right] \times 10^{-6} \qquad \text{(A.17)}$$

$$= 3526 \text{ kN} > M_{Ed} \therefore \text{Ok!}$$

3. Design shear resistance

$$V_{pl,a,Rd} = A_v \frac{f_{yd}}{\sqrt{3}}$$

$$V_{pl,a,Rd} = \left[ 668 \times 18 \times \frac{460}{\sqrt{3}} \right] \times 10^{-3} = 2483.7 \text{ kN} > V_{Ed} \therefore \text{Ok!} \quad \text{(A.18)}$$

Shear buckling check

$$\frac{h_w}{t_w} \leq 72\varepsilon$$

$$\frac{668}{14} = 47.7 \leq \left( 72 \times \sqrt{\frac{235}{460}} \right) = 51.5 \therefore \text{Ok!} \qquad \text{(A.19)}$$

4. Shear connection
   Refer to Figure A.5. Stud failure

$$P_{Rd} = \frac{0.8 f_u \pi d^2}{4\gamma_v} = 0.16 f_u \pi d^2$$

$$P_{Rd} = \left( 0.16 \times 500 \times \pi 19^2 \right) \times 10^{-3} = 90.7 \text{ kN} \qquad \text{(A.20)}$$

## CONCRETE FAILURE

$$P_{Rd} = \frac{0.29\alpha d^2 \sqrt{f_{ck}E_{cm}}}{\gamma_v} = 0.232\alpha d^2 \sqrt{f_{ck}E_{cm}}$$

$$\left(\frac{h_{sc}}{d}+1\right) = \left(\frac{95}{19}+1\right) = 5 > 4 \therefore \alpha = 1$$

$$P_{Rd} = 0.232 \times 1 \times 19^2 \sqrt{55 \times 31000} = 109 \text{ kN}$$

$$\therefore P_{Rd} = 91.7 \tag{A.21}$$

Reduction factor for parallel profile steel sheeting

$$k_l = 0.6\frac{b_0}{h_p}\left(\frac{h_{sc}}{h_p}-1\right) \leq 1.0$$

$$k_l = 0.6 \times \frac{150}{54} \times \left(\frac{95}{54}-1\right) = 1.2 \leq 1.0$$

$$\therefore k_l = 1.0$$

$$\therefore P_{Rd} = 91.7 \tag{A.22}$$

HPCB is designed to full shear connection

$$\eta = \frac{N_{Rd}}{\min\left(N_{c,f}, N_{pl,a}\right)}$$

$$1 = \frac{N_{Rd}}{7382.8}$$

$$N_{Rd} = 7382.8 \text{ kN}$$

Row of shear studs required

$$no = \frac{N_{Rd}}{R_{Rd}}$$

$$no = \frac{7382.8}{91.7 \times 2} \approx 41nos$$

Spacing between shear studs

$$s = \frac{L_{eff}/2}{n-1}$$

$$s = \frac{7200/2}{41-1} = 90 \text{ mm}$$

$$\therefore s = 90 \text{ mm} > \text{min spacing} \therefore \text{Ok!}$$

The design bending moment and shear resistance of the primary and secondary beams are summarized in Table A.1. The geometrical details of the primary and secondary beams are also tabulated in Table A.2 for reference.

TABLE A.1　Summary table for design bending moment and shear resistance

|  | Moment (kNm) | | Shear (kN) | | |
|---|---|---|---|---|---|
|  | $M_{Ed}$ | $M_{Rd}$ | $V_{Ed}$ | $V_{Rd}$ | PNA (mm) |
| Secondary HPCB | 1753 | 1779 | 584 | 1258 | 54.5 |
| Primary HPCB | 3214 | 3526 | 2365 | 2483 | 116 |

TABLE A.2　Summary table of section details

|  | Top Flange (mm) | | Web (mm) | | Bottom Flange (mm) | | Shear Connections | |
|---|---|---|---|---|---|---|---|---|
|  | Width | Thickness | Height | Thickness | Width | Thickness | No/row | Spacing(mm) |
| Secondary HPCB | 140 | 10 | 474 | 10 | 170 | 16 | 2 | 200 |
| Primary HPCB | 140 | 14 | 668 | 14 | 230 | 18 | 2 | 80 |

## A.3 SERVICEABILITY LIMIT STATE

For simplicity, only the deflection and vibration of the secondary beam are shown here as an example.

General criteria

$$A_a\left(h_t + h_a - h_c - h_s\right) > \frac{b_{\text{eff}}h_c^2}{2n}$$

$$8660\left(180 + 500 - 126 - 216\right) > \frac{2400 \times 126^2}{2 \times \dfrac{210000}{38000}}$$

$$2.99 \times 10^6 < 3.45 \times 10^6 \therefore \text{ENA in concrete} \quad (\text{A.23})$$

Position of neutral axis

$$X_{\text{ENA}} = \frac{A_a\left(h_t + h_a - h_c - h_s\right) + \dfrac{b_{\text{eff}}h_c^2}{2n}}{A_a + \dfrac{b_{\text{eff}}h_c}{n}}$$

$$X_{\text{ENA}} = \frac{8860\left(180 + 500 - 126 - 216\right) + \dfrac{2400 \times 126^2}{2 \times \dfrac{210000}{38000}}}{8860 + \dfrac{2400 \times 126}{2 \times \dfrac{210000}{38000}}}$$

$$= 118.9 \text{ mm} \quad (\text{A.24})$$

Effective second moment of area

$$I_{eff} = I_a + \frac{A_a \left(h_c + 2h_d + h_a\right)^2}{4\left(1 + nr\right)} + \frac{b_{eff} h_c^3}{12n}$$

$$I_a = \left[ \frac{b_{f1} t_{f1}^3}{12} + b_{f1} t_{f1} \bullet \left(h_s - \overline{x}_{f1}\right)^2 + \frac{t_w h_w^3}{12} + t_w h_w \bullet \left(h_s - \overline{x}_w\right)^2 + \frac{b_{f2} t_{f2}^3}{12} \right.$$

$$\left. + b_{f2} t_{f2} \bullet \left(h_s - \overline{x}_{f2}\right)^2 \right]$$

$$I_a = \frac{120 \times 10^3}{12} + 120 \times 10 \times \left(500 - 216 - \frac{10}{2}\right)^2 + \frac{10 \times 474^3}{12}$$

$$+ 10 \times 474 \times \left(16 + \frac{474}{2} - 216\right)^2 + \frac{170 \times 16^3}{12}$$

$$+ 170 \times 16 \times \left(216 - \frac{16}{2}\right) = 3.22 \times 10^{82}$$

$$I_{eff} = 3.22 \times 10^8 + \frac{8860\left(126 + 2\left(54\right) + 500\right)^2}{4\left(1 + \dfrac{210000}{38000} \bullet \dfrac{8860}{2400 \times 126}\right)} + \frac{2400 \times 126^3}{12\left(\dfrac{210000}{38000}\right)}$$

$$= 1.42 \times 10^9 \qquad\qquad (A.25)$$

Deflection for uniform distributed load (UDL)

$$\delta = \frac{5wL^4}{384 E I_{eff}}$$

$$\delta = \frac{5 \times 25 \times 12000^4}{384 \times 210000 \times 1.42 \times 10^9} = 22.6 \text{ mm} \qquad (A.26)$$

Vibration response check
Natural frequency

$$f_n = \frac{18}{\sqrt{\delta_w}} > 4 \text{ Hz}$$

$$\delta_w = \frac{5wL^4}{384EI_{\text{eff}}}$$

$w$ = full permanent load + 10% varaiable action = 14.5 kN/m

$$\delta_w = \frac{5(14.5)(12000)^4}{384 \times 210000 \times 1.42 \times 10^9} = 13.1 \text{ mm}$$

$$f_n = \frac{18}{\sqrt{13.1}} = 4.9$$

$$(A.27)$$

Modal mass

$$L_{\text{eff}} = 1.09(1.10)^{n_y-1}\left(\frac{EI_b}{mbf_0^2}\right)^{\frac{1}{4}}$$

$$L_{\text{eff}} = 1.09(1.10)^{5-1}\left(\frac{(210000)1.377 \times 10^7}{14.5(2.4)(4.9)^2}\right)^{\frac{1}{4}} = 15.25 \text{ m} \qquad (A.28)$$

$$S = \eta(1.15)^{n_x-1}\left(\frac{EI_b}{mbf_0^2}\right)^{\frac{1}{4}}$$

$$\eta = 0.21 \times 4.9 - 0.55 = 0.48$$

$$S = 0.48(1.15)^{2-1}\left(\frac{210000(1.377 \times 10^7)}{14.5(2.4)(4.9)^2}\right)^{\frac{1}{4}} = 5.5 \text{ m} \qquad (A.29)$$

$$M = mL_{\text{eff}}S$$

$$M = \frac{14.5 \times 100}{2.4}(15.25)(5.5) = 50531.79 \text{ kg} \qquad (A.30)$$

Floor response

$$a_{w,rms} = \mu_e \mu_r \frac{0.1Q}{2\sqrt{2}M\xi} W\rho$$

$Q = 746$ N based on average mass of 76 kg

$\xi = 4.68\%$

$W = 1$

$\mu_e = \mu_r = 1$

$$\rho = 1 - e^{\left(\frac{-2\pi\xi L_p f_p}{v}\right)} = 1 - e^{\left(\frac{-2\pi(0.0468)(15)(2)}{1.52}\right)} = 0.99$$

$$a_{w,rms} = (1)(1)\frac{0.1(746)}{2\sqrt{2}(50531.79)(0.0468)}(1)(0.99) = 0.011 \quad (A.31)$$

Response factor

$$R = \frac{a_{w,rms}}{0.005}$$

$$R = \frac{0.011}{0.005} = 2.21 \quad (A.32)$$

Refer to Table A.3 for the vibration response limit. This HPCB design is adequate for use as office or production buildings, but not suitable for use as semi-conductor industrial buildings.

TABLE A.3   Vibration response limit

| Building Type | Min Vibration Requirements | Max R |
|---|---|---|
| Semi-conductor industrial (high end) | VC-D | 0.625 |
| Semi-conductor industrial (low end) | VC-B | 0.25 |
| Office | ISO for Office | 4 |
| Production | ISO for Workshop | 8 |

# Case Study

*Feasibility, Productivity, Cost and Carbon Analysis*

## B.1 ADVANCED RC SOLUTION

The next generation industrial building for this case study is a typical heavily reinforced precast and cast in-situ RC structural framing system using multiple secondary single "tee" slab-beams connected to main beams, as shown in Figures B.1. The secondary beams are spaced out at 2,000 mm centre to centre. The common concrete classes for this RC solution are C70/85 for all columns and C40/50 for precast and cast in-situ beams and slab. The typical sizes of the columns are 900 mm by 1,200m, 1000 mm by 1,500 mm, 1,200 mm by 1,200 mm, and 1,200 mm by 1,800 mm and the average reinforcement ratio is 3%. While the sizes for the single "tee" slab-beam vary from 600 mm to 1,400 mm in depth and 350 mm to 450 mm in width. The main beam has a standard width size of 2,000 mm and a varying depth ranging from 1,000 mm to 1,400 mm. The sizes of the structural elements are

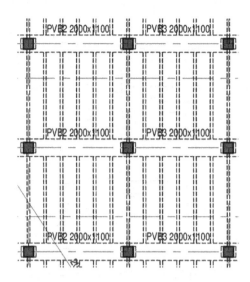

FIGURE B.1    Part floor plan for $CO_2e$ analysis.

TABLE B.1    Original design of the advanced reinforced concrete structural system

| Column | | Primary Beam | | Secondary Beam | |
|---|---|---|---|---|---|
| Width | Depth | Width | Depth | Width | Depth |
| 900 | 1,200 | 1,200 | 1,200 | 600 | 350 |
| 1,000 | 1,500 | 2,000 | 1,100 | 700 | 350 |
| 1,200 | 1,200 | 2,000 | 1,200 | 850 | 350 |
| 1,200 | 1,800 | 2,000 | 1,400 | 1,000 | 400 |
| | | | | 1,200 | 1,400 |
| | | | | 1,400 | 1,400 |

listed in Table B.1. The typical reinforcement details of main and secondary beams are shown in Figures B.2 and B.3. The typical reinforcement details of columns are presented in Figure B.4.

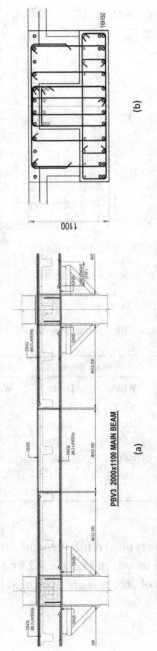

FIGURE B.2  Longitudinal and cross-sectional views of the primary beam. (a) Longitudinal section; and (b) Cross-section.

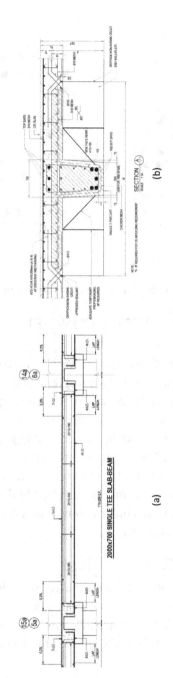

FIGURE B.3　Longitudinal and cross-sectional views of the secondary beam. (a) Longitudinal section; and (b) Cross-section.

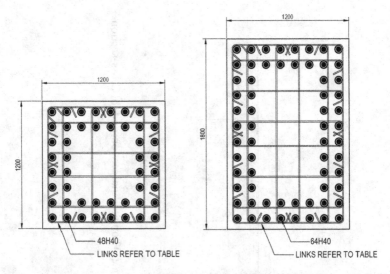

FIGURE B.4   Cross-sections of the 1.2 m ×1.2 m and 1.2 m ×1.8 m columns.

## B.2  CONVENTIONAL STEEL CONCRETE COMPOSITE STRUCTURAL SYSTEM

The structural layout of conventional composite construction for the 12 m by 12 m grid is as shown in Figure A.1. The secondary beams are spaced out at 2.4 m, centre to centre, apart, and with a composite slab thickness of 250 mm. The size of the primary beams used is standard UB of 914 × 419 × 388 kg/m, and the size of secondary beams is 533 × 210 × 92 kg/m. It shall be mentioned that the size of the structural members chosen has a utilisation rate of 0.92. For a better compression of embodied carbon, the column of this composite construction is adopting RC columns with concrete class of C70/80.

## B.3  HPCB STRUCTURAL SYSTEM

Similar to the conventional composite construction, the HPCB construction has the same structural layout with secondary beams spaced out at 2.4 m, centre to centre, and a composite slab thickness of 180 mm. In similar fashion, the beam elements are designed to have a utilisation rate of 0.92 for fair comparison. For a better illustration of the connection of primary HPCB, the longitudinal sections and cross sectional details are in shown Figures B.5 and B.6 respectively. The longitudinal sections of secondary HPCB and its cross sectional details are presented in Figures B.7 and B.8.

FIGURE B.5   Longitudinal section of the primary HPCB.

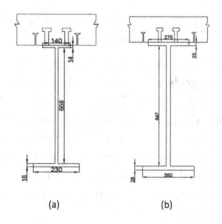

(a)                          (b)

FIGURE B.6   Cross-sectional details of the primary HPCB. (a) Sagging portion; and (b) Hogging portion.

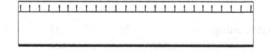

FIGURE B.7   Longitudinal section of secondary HPCB.

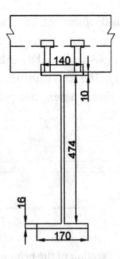

FIGURE B.8   Cross-sectional details of the secondary HPCB.

## B.4 PRODUCTIVITY AND CARBON ANALYSIS

### B.4.1 Carbon Analysis - Flooring Comparison

This section discusses the carbon emission of the three systems, focusing on the 4 grids of 12 m by 12 m layout of slabs and beams involved, excluding columns. The weight of each system is shown in Table B.2 and the carbon emission is shown in Figure B.9. As expected, the proposed HPCB solution returns the lowest carbon emission among the three systems, with 456 kg/GFA of $CO_2$ equivalent, followed by conventional composite construction with 547 kg/GFA and lastly the conventional reinforced concrete system with 650 kg/GFA.

TABLE B.2   Weight comparison (kg per GFA)

|  | **Conventional RC** | **Conventional Composite** | **HPB** |
|---|---|---|---|
| Steel |  | 91 | 57 |
| Profile Steel Sheeting |  | 11 | 11 |
| Concrete | 1,362 | 600 | 432 |
| Reinforcement | 115 | 78 | 56 |
| Shear Links | 41 |  |  |

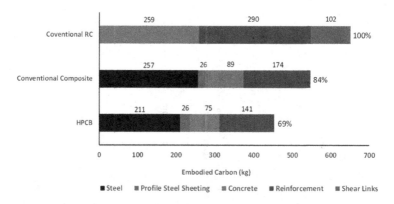

FIGURE B.9    CO$_2$e comparison for the flooring systems.

1517.7 kg/GFA of materials are required for the RC solution, and 90% of it is contributed by concrete. The conventional composite construction reduces 56% of the concrete used by replacing with structural steel sections. This reduces CO$_2$e produce by concrete and reinforcement from 259 kg/GFA to 290 kg/GFA and 174 kg/GFA to 141 kg/GFA respectively. This is CO$_2$e is further enhanced by HPCB with the use of HPGC to reduce the size of concrete slab, and this reduces 28% of concrete and reinforcement required for slab. This resulted a further CO$_2$e saving of 16.2% from concrete and 28% from reinforcement. As redundant weight is reduced from the UB steel section through structural optimisation of HPCB, resulted in 38% saving in weight for structural steel and 31% reduction in CO$_2$e. Furthermore, the overall headroom was reduced from the conventional RC system and composite construction with the used of HPCB system, as shown in Figure B.10.

## B.4.2  Carbon Analysis - Super Structure

The previous sections have separately discussed the carbon emission of the flooring system, without consideration of the columns. As highlighted in the previous section, carbon emissions can be reduced through the use of high-performance materials

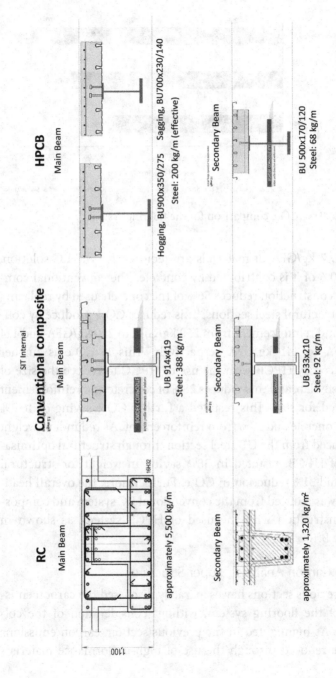

FIGURE B.10  Comparison among as-built RC, conventional composite and HPCB designs.

and optimisation of structural design to use lesser materials to achieve the same capacity. Therefore, the embodied carbon in this section is analysing the 4 grids of 12 m by 12, with the inclusion of columns. As explained in the earlier section, both HPCB and conventional composite are adopting RC columns. The foundation system has been excluded from this study as it involves too many uncertainties and simplified assumptions may alter the comparison.

The comparison of embodied carbon between the systems is shown in Figure B.11, and the weight of each system is shown in Table B.3. The reinforced concrete system has the highest embodied carbon, followed by the conventional composite system and lastly the proposed pre-engineered composite system, with 967 kg/GFA, 775 kg/GFA, and 574 kg respectively. It is evident that the HPCB is able to reduce the embodied carbon of RC system by approximately of 40%, from 967 kg per GFA to 574 kg per GFA. It is due to the lighter loading onto the structural system as the weight of the structure has been significantly reduced from 2237 kg/GFA to 848 kg/GFA. This weight reduction is estimated to be 62%. The weight reduction in the columns alone is 27.5% from RC to conventional composite construction and 59.4% from RC system to HPCB solution.

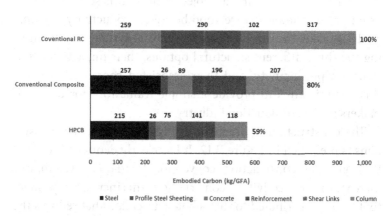

FIGURE B.11   $CO_2$e comparison for structural system.

TABLE B.3    Weight comparison for the superstructure (kg per GFA)

|  | Conventional RC | Conventional Composite | HPB |
|---|---|---|---|
| Steel |  | 91 | 57 |
| Profile Steel Sheeting |  | 105 | 11 |
| Concrete | 1,362 | 600 | 432 |
| Reinforcement | 115 | 78 | 56 |
| Shear Links | 41 |  |  |
| Columns | 720 | 521 | 292 |

## B.4.3 Productivity and Cost Analysis

Within the domain of construction and civil engineering, the process of choosing a suitable structural system holds paramount importance as it possesses the potential to exert a substantial impact on the result of a given project. The structural systems of a building or infrastructure play a crucial role in providing the necessary support and stability, as well as ensuring its long-term durability and overall performance. Within the decision-making process, there are numerous elements that necessitate consideration. However, two factors emerge as particularly significant: productivity and cost. The interconnection between these two aspects is complex and interdependent, requiring project managers and engineers to carefully navigate and maintain equilibrium in order to attain favourable project results. This section aims to examine the complex correlation between productivity and cost analysis inside the three different structural systems. By examining the three different structural options, their impacts on construction productivity, and their associated construction costs, this analysis aims to provide a simple framework for decision-makers in the construction industry.

The construction productivity of the different structural systems is presented in Figure B.12. It has clearly demonstrated that the composite construction, conventional or HPCB system, can increase the productivity of construction. This increase in productivity can be attributed to the off-site fabrication that reduces the

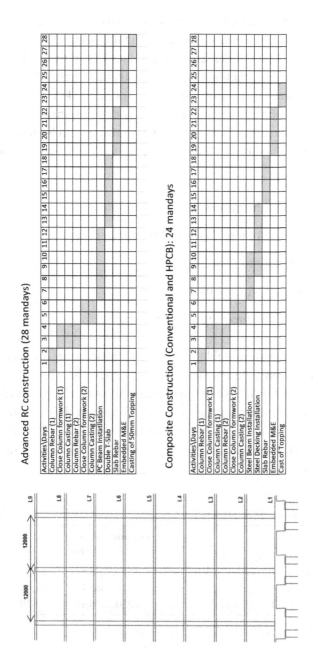

FIGURE B.12   Productivity comparison between advanced RC and composite construction.

in-situ work preparation. In addition, due to the reduction in mass of each structural element, in particular the beam element, this has also significantly increased the productivity in on-site installation. It is worth noting that the construction productivity in these two figures is based on a single level of 24 by 24 m. The actual productivity would, therefore, increase by a minimum of 12%.

The cost comparison of the three structural system is presented in Table B.4. To simplify the analysis, the following assumptions are made:

1. 8 Floors with option system + CIS L1 floor

2. Typical 12x12m grid (with 24 m × 24 m part plan)

3. Typical floor to floor height: 11 m

4. Loadings: SDL 2.9 kPa, LL 20 kPa

5. Standard fire proofing for 2 hr rating

As can be seen from Table B.4, the conventional steel concrete composite beam is 15.8% more expensive than the advanced RC solution, which is one of the main reasons why industrial buildings are predominantly RC structures in Singapore. By optimizing the geometry and reducing the redundant weight, HPCB requires less input of materials and eventually achieve 7.9% cost saving at the time of analysis.

TABLE B.4   Cost comparison

| Scheme | Estimated cost SGD (per 24 m × 2 m) | % |
|---|---|---|
| CIS Column + Precast beam & tee slab | $3,183,094.68 | 0 |
| CIS Column + Composite system | $3,687,083.17 | +15.83% |
| CIS Column + HPCB system | $2,931,403.49 | −7.91% |

Accurate at the time of production

# Index

Pages in *italics* refer to figures and pages in **bold** refer to tables.

Printed in the United States
by Baker & Taylor Publisher Services